EAI/Springer Innovations in Communication and Computing

Series Editor
Imrich Chlamtac, European Alliance for Innovation
Ghent, Belgium

The impact of information technologies is creating a new world yet not fully understood. The extent and speed of economic, life style and social changes already perceived in everyday life is hard to estimate without understanding the technological driving forces behind it. This series presents contributed volumes featuring the latest research and development in the various information engineering technologies that play a key role in this process. The range of topics, focusing primarily on communications and computing engineering include, but are not limited to, wireless networks; mobile communication; design and learning; gaming; interaction; e-health and pervasive healthcare; energy management; smart grids; internet of things; cognitive radio networks; computation; cloud computing; ubiquitous connectivity, and in mode general smart living, smart cities, Internet of Things and more. The series publishes a combination of expanded papers selected from hosted and sponsored European Alliance for Innovation (EAI) conferences that present cutting edge, global research as well as provide new perspectives on traditional related engineering fields. This content, complemented with open calls for contribution of book titles and individual chapters, together maintain Springer's and EAI's high standards of academic excellence. The audience for the books consists of researchers, industry professionals, advanced level students as well as practitioners in related fields of activity include information and communication specialists, security experts, economists, urban planners, doctors, and in general representatives in all those walks of life affected ad contributing to the information revolution.

Indexing: This series is indexed in Scopus, Ei Compendex, and zbMATH.

About EAI - EAI is a grassroots member organization initiated through cooperation between businesses, public, private and government organizations to address the global challenges of Europe's future competitiveness and link the European Research community with its counterparts around the globe. EAI reaches out to hundreds of thousands of individual subscribers on all continents and collaborates with an institutional member base including Fortune 500 companies, government organizations, and educational institutions, provide a free research and innovation platform. Through its open free membership model EAI promotes a new research and innovation culture based on collaboration, connectivity and recognition of excellence by community.

More information about this series at https://link.springer.com/bookseries/15427

Sanjay Chakraborty • SK Hafizul Islam
Debabrata Samanta

Data Classification and Incremental Clustering in Data Mining and Machine Learning

Springer

Sanjay Chakraborty
JIS University
Dum Dum Cantonment, India

SK Hafizul Islam
Indian Institute of Information
Technology Kalyani
West Bengal, India

Debabrata Samanta
Department of Computer Science
CHRIST (Deemed to be University)
Bangalore, India

ISSN 2522-8595   ISSN 2522-8609   (electronic)
EAI/Springer Innovations in Communication and Computing
ISBN 978-3-030-93090-5   ISBN 978-3-030-93088-2   (eBook)
https://doi.org/10.1007/978-3-030-93088-2

This Springer imprint is published by the registered company Springer Nature Switzerland AG
The registered company address is: Gewerbestrasse 11, 6330 Cham, Switzerland

To my parents, wife Lopamudra Dey and son Arohan Chakraborty for their love and inspiration.

Sanjay Chakraborty

To my son Enayat Rabbi

Dr SK Hafizul Islam

To my parents Dulal Chandra Samanta and Ambujini Samanta, my elder sister Tanusree Samanta, and daughter Aditri Samanta

Dr Debabrata Samanta

Preface

This book can be recommended for any basic courses in data mining and machine learning (ML). It can also be a very useful book for those conducting research in supervised and unsupervised learning techniques. This book can be the first step in understanding data mining, machine learning and its applications in real-life data, giving special attention to classification and clustering techniques. It offers a crisp and compact discussion on data mining and machine learning techniques with some case studies and real-life examples. It discusses certain short real-life projects that can help students get a clear idea about application of topics, and, finally, it discusses a set of industry-based questions and solutions to help prepare a candidate for the position of data scientist or data engineer.

Chapter 1 expresses data mining as the process of extracting useful hidden patterns from a huge amount of data stored across numerous heterogeneous resources. Data mining is quite useful for corporate executives to make strategic judgements after analysing the hidden truth of data. It is one of the steps in the process of knowledge generation. A data warehouse, a database server, a data mining engine, a pattern analysis module and a graphical user interface make up a data mining system. Mining common patterns and association rule learning with analysis, as well as sequence analysis, are examples of data mining approaches. Data mining is a technique that may be used on top of many types of intelligent data stores, such as data warehouses, to perform analysis and take strategic decisions. A data mining system in a huge database faces a number of concerns and challenges. It is an excellent environment for data scientists and developers to work.

Machine learning is a subset of AI. It is a field aimed at gathering computer programmes capable of performing intelligent actions based on prior facts or experiences. Most of us utilise machine learning on a daily basis when we use Netflix, YouTube and Spotify recommendation algorithms, as well as Google and Yahoo search engines and voice assistants like Google Home and Amazon Alexa. All of the data is labelled, and algorithms learn to anticipate the output from the input. The algorithms learn from the data's underlying structure, which is unlabelled. Because some data is labelled but not all is, a combination of supervised and unsupervised techniques can be used, as expressed in Chap. 2.

Chapter 3 shows that an unknown example can be classified with the most common class among K closest examples using a Supervised Learning–Based Data Classification and Incremental Clustering. "Tell me who your neighbours are, and it will tell you who you are", says the KNN classifier. The Supervised Learning–Based Data Classification and Incremental Clustering is a simple yet effective algorithm that has a wide range of applications in computer vision, pattern identification, optical character recognition, facial recognition, genetic pattern recognition and other domains. It is also known as a slacker learner because it waits until the last minute to build a model to classify a given test tuple.

Clustering is studied using data modelling, which is based on mathematics, statistics and numerical analysis. Clusters in machine learning allude to hidden patterns, and clusters are discovered through unsupervised learning, with the final system being a data idea. As a result, clustering is the ad hoc finding of a previously unknown data idea. The fact that data mining deals with enormous databases increases the computing requirements of clustering analysis. Data mining clustering techniques that are both powerful and broadly applicable have arisen as a result of these issues. In some applications, clustering is also known as data segmentation since it divides big data sets into categories based on their commonalities. Outliers (values that are "far away" from any cluster) can be more fascinating than usual examples, hence clustering can be used to find them. The detection of credit card fraud and the surveillance of illegal online commercial activity are examples of outlier detection applications, express in Chap. 4.

Chapter 5 focuses on the process of grouping new incoming or incremental data into classes or clusters, which is known as incremental clustering. It mostly clusters new data that arrives at random into a comparable group of clusters. The traditional K-means and DBSCAN clustering methods are inefficient for handling big dynamic databases since they simply run their algorithms again and again for each change in the incremental database, taking a long time to effectively cluster the new arriving data. It not only takes too much time, but it has also been discovered that using the existing algorithm for updated databases on a regular basis may be too expensive. As a result, the current K-means clustering approach is ineffective in a dynamic setting. To address these issues, we've implemented incremental versions of K-means and DBSCAN in our research.

The most expressive research and application domains are data mining and machine learning. All real-time applications rely on data mining and machine learning, either directly or indirectly. There are numerous applications, including data analysis in finance, retail, telecommunications, biological data analysis and further scientific uses, and intruder detection, as shown in Chap. 6.

Chapter 7 projects that, when it comes to high-dimensional data processing, scientists and analysts in the fields of machine learning (ML) and data mining have a challenge. Variable selection is an excellent way to deal with this problem since it eliminates redundant and unneeded data, reducing calculation time, improving learning accuracy and making the learning approach or data easier to understand. We present numerous regularly used variable selection evaluation metrics in this chapter before surveying supervised, unsupervised and semi-supervised variable selection strategies that are commonly utilised in ML tasks such as classification and clustering. Finally, the issues associated with variable selection are addressed. Variant selection is an important problem in machine learning and pattern recognition, and several approaches have been proposed. Using public-domain datasets, we examine the performance of various variable selection approaches in this study. We measured the number of decreased variants as well as the rise in learning assessment using the specified variable selection strategies, and then assessed and compared each approach using these metrics.

During the variation selection method, a subset of accessible variants data is chosen for the learning algorithms. This comprises the most crucial, which has the fewest parameters and has the most impact on learner accuracy. The advantage of variant selection is that vital information about a particular variant is preserved, but if just a small number of variants are required and the original variants are exceedingly diverse, there is a risk of information being lost because certain variants must be discarded. On the other hand, dimensional reduction, which is also based on variant extraction, reduces the size of the variant space without sacrificing information from the original variant space, as expressed in Chap. 8.

Dum Dum Cantonment, India   Sanjay Chakraborty
West Bengal, India   SK Hafizul Islam
Bangalore, India   Debabrata Samanta

Acknowledgement

We express our great pleasure, sincere thanks and gratitude to the people who significantly helped, contributed and supported the completion of this book. We are sincerely thankful to Prof. G. P. Biswas, Emeritus Professor, Department of Computer Science and Engineering, Indian Institute of Technology (Indian School of Mines) Dhanbad, Jharkhand, India, for his encouragement, support, guidance, advice and suggestions to complete this book. Our sincere thanks to Fr. Benny Thomas, Professor, Department of Computer Science and Engineering, CHRIST (Deemed to be University), Bengaluru, Karnataka India, and Dr Arup Kumar Pal, Associate Professor, Department of Computer Science and Engineering, Indian Institute of Technology (Indian School of Mines) Dhanbad, Jharkhand, India, for their continuous support, advice and cordial guidance from the beginning to the completion of this book.

We would also like to express our honest appreciation to our colleagues at the JIS University, India, Indian Institute of Information Technology Kalyani, and CHRIST (Deemed to be University), Bengaluru, Karnataka India, for their guidance and support.

We also thank all the authors who have contributed some chapters to this book. This book would not have been possible without their contribution.

We are also very thankful to the reviewers for reviewing the book chapters. This book would not have been possible without their continuous support and commitment towards completing the review on time.

To complete this book, the entire staff at Springer extended their kind cooperation, timely response, expert comments and guidance, and we are very thankful to them.

Finally, we sincerely express our special and heartfelt respect, gratitude and gratefulness to our family members and parents for their endless support and blessings.

Department of Computer Science and Engineering   Sanjay Chakraborty
JIS University
Dum Dum Cantonment, India

Department of Computer Science and Engineering
Indian Institute of Information Technology
Kalyani, West Bengal, India

SK Hafizul Islam

Department of Computer Science
CHRIST (Deemed to be University)
Bengaluru, Karnataka, India

Debabrata Samanta

Contents

About the Authors

Sanjay Chakraborty is currently an assistant professor in the Department of Computer Science and Engineering, JIS University, Kolkata, India. He obtained his BTech from West Bengal University of Technology, India, in information technology in the year 2009. He obtained his MTech degree from the National Institute of Technology, Raipur, India, in 2011. He submitted his PhD thesis at AKCSIT, University of Calcutta, in 2020. Mr. Chakraborty is the recipient of the University Silver Medal from NIT Raipur in 2011 for ranking first-class second in MTech. He has more than 10 years of teaching and research experience. He has published over 50 research papers in various international journals, conferences and book chapters. He is the author of one book on ML-based brain-computer interfacing published by Lap Lambert, Germany. Mr. Chakraborty attended many national and international conferences in India and abroad. His research interests include data mining, machine learning and quantum computing. He is a professional member of IAENG and UACEE. Mr. Chakraborty is an active member of the board of reviewers of various international journals and conferences. He is the recipient of "INNOVATION AWARD" for outstanding achievement in the field of innovation by Techno India Institution's Innovation Council 2019. Mr. Chakraborty is also the recipient of "IEEE Young Professional Best Paper Award" in 2017. He has also achieved the top five best paper recognition by Ain Shams Engineering Journal, Elsevier. He is a reviewer of various reputed journals and conferences, such as *IEEE Transactions* and *Nature*.

SK Hafizul Islam received his M.Sc. in Applied Mathematics from Vidyasagar University, Midnapore, India, in 2006, and his M.Tech. in Computer Application and Ph.D. in Computer Science and Engineering in 2009 and 2013, respectively, from the Indian Institute of Technology [IIT (ISM)] Dhanbad, Jharkhand, India, under the INSPIRE Fellowship PhD Program (funded by Department of Science and Technology, Government of India). He is currently an Assistant Professor in the Department of Computer Science and Engineering, Indian Institute of Information Technology Kalyani (IIIT Kalyani), West Bengal, India. Before joining the IIIT Kalyani, he was an Assistant Professor in the Department of Computer Science and

Information Systems, Birla Institute of Technology and Science, Pilani (BITS Pilani), Rajasthan, India. He has more than 10 years of teaching and 12 years of research experience. He has authored or co-authored 125 research papers in internationally reputed journals and conference proceedings. His research interests include cryptography, information security, WSNs, IoT and cloud computing. Dr Islam is an Associate Editor for "IEEE Journal of Biomedical and Health Informatics", "IEEE Transactions on Intelligent Transportation Systems", "IEEE Access", "International Journal of Communication Systems (Wiley)", "Telecommunication Systems (Springer)", "IET Wireless Sensor Systems", "Security and Privacy (Wiley)", "Array- Journal (Elsevier)", and Technical Committee Member, Computer Communications (Elsevier). He has been a reviewer for many reputed international journals and conferences. He has been recipient of the University Gold Medal, the S. D. Singha Memorial Endowment Gold Medal and the Sabitri Parya Memorial Endowment Gold Medal of Vidyasagar University, in 2006. He has also been recipient of the University Gold Medal of IIT(ISM) Dhanbad in 2009 and the OPERA award of BITS Pilani in 2015. Dr Islam is a senior member of the IEEE and a member of the ACM.

Debabrata Samanta is presently working as Assistant Professor, Department of Computer Science, CHRIST (Deemed to be University), Bangalore, India. He obtained his Bachelors in Physics (Honors), from Calcutta University; Kolkata, India. He obtained his MCA, from the Academy of Technology, under WBUT, West Bengal. He obtained his PhD in Computer Science and Engg. from National Institute of Technology, Durgapur, India, in the area of SAR Image Processing. He is keenly interested in Interdisciplinary Research & Development and has experience spanning fields of SAR Image Analysis, Video surveillance, Heuristic algorithm for Image Classification, Deep Learning Framework for Detection and Classification, Blockchain, Statistical Modelling, Wireless Adhoc Network, Natural Language Processing, V2I Communication. He has successfully completed six Consultancy Projects. He has received funding 5250 USD under Open Access, Publication fund. He has received funding under International Travel Support Scheme in 2019 for attending conference in Thailand. He has received Travel Grant for speaker in Conference, Seminar etc for two years from July, 2019. He is the owner of 21 Patents (3 Design Indian Patent and 2 Australian patent Granted, 16 Indian Patents published) and 2 copyright. He has authored and coauthored over 185 research papers in international journal (SCI/SCIE/ESCI/Scopus) and conferences including IEEE, Springer and Elsevier Conference proceeding. He has received "Scholastic Award" at 2nd International conference on Computer Science and IT application, CSIT-2011, Delhi, India. He is a co-author of 12 books and the co-editor of 11 books, available for sale on Amazon and Flipkart. He has presented various papers at International conferences and received Best Paper awards. He has author and co-authored of 08 Book Chapters. He also serves as acquisition editor for Springer, Wiley, CRC, Scrivener Publishing LLC, Beverly, USA and Elsevier. He is a Professional IEEE Member, an Associate Life Member of Computer Society Of India (CSI) and a Life Member of Indian Society for Technical Education (ISTE).

He is a Convener, Keynote speaker, Session chair, Co-chair, Publicity chair, Publication chair, Advisory Board, Technical Program Committee members in many prestigious International and National conferences. He was invited speaker at several Institutions.

Chapter 1
Introduction to Data Mining and Knowledge Discovery

1.1 Introduction

This chapter introduces the concepts of data mining, incremental data mining, and its applications and limitations. Data mining is an essential tool used in various research fields now a day. "Data abounds, but information is scarce" – data mining emerges to address this shortcoming. Data mining is a new and exciting field that aims to extract meaningful information from massive amounts of data. That is why it is also known as "data-driven knowledge discovery". It can also refer to the process of extracting knowledge from enormous amounts of data, similar to how gold is extracted from rocks in gold mining [1, 2]. Data mining is the process of discovering previously undiscovered, valid patterns and relationships in big data sets using advanced data analysis techniques. These technologies include statistical models, mathematical algorithms and machine learning methodologies. The following "Fig. 1.1" shows how data mining works.

Data mining software analyzes data and can reveal valuable data trends, which can help with company strategy, knowledge bases, and scientific and medical research. Many various types of databases, including multimedia databases, can be mined for data patterns and so on [3–4].

1.2 Architecture of Data Mining System

From the basic view of data mining functionality, a typical data mining system can be developed. A typical data mining system's design is depicted in "Fig. 1.2". It consists of six modules; they are given below:

© The Author(s), under exclusive license to Springer Nature Switzerland AG 2022
S. Chakraborty et al., *Data Classification and Incremental Clustering in Data Mining and Machine Learning*, EAI/Springer Innovations in Communication and Computing, https://doi.org/10.1007/978-3-030-93088-2_1

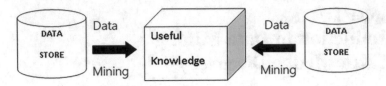

Fig. 1.1 Diagram of knowledge extraction from a huge database

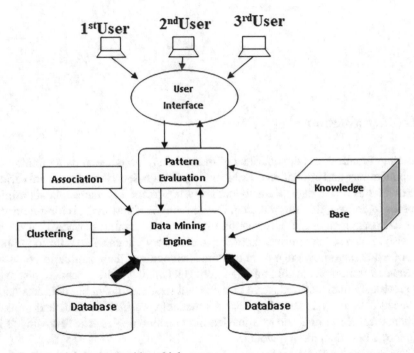

Fig. 1.2 Data mining system with multiple user

  (i) **User interface**: This module allows users to interface with the data min-
      ing system.
 (ii) **Pattern evaluation**: This module connects with the data mining modules and
      employs interestingness measurements [5].
(iii) **Data mining engine**: This vital component consists of a set of functional mod-
      ules for tasks like characterisation, correlation, cluster analysis and outlier
      analysis, among others.
 (iv) **Knowledge base**: This module is used to direct the search or assess the useful-
      ness of the patterns found.
  (v) **Data repository**: This is one or set of large databases.

1.2.1   Knowledge Discovery Process (KDD)

Data mining process is basically based on knowledge discovery process (KDD). The core process in *KDD* is known as data mining. KDD is a non-trivial process of identifying valid, novel, potentially useful and understandable patterns and relation-ships in data [6].

Steps of KDD process are as follows:

- Gain a thorough understanding of the application domain, relevant prior knowl-edge, and the end user's objectives.
- Create a data set that will be used as a target.
- Eliminate noise and outliers; deal with missing data and unlabelled data.
- Identify essential attributes that can be used to describe data better.
- Pick a data mining project.
- Make a decision on the data mining algorithm.
- Data mining is number seven.
- Analyze the patterns you've discovered.
- Compile the information you've gathered.

Data mining has two high-level goals: prediction and prediction of unknown or future values of specified variables. The types of databases mined, the kinds of knowledge mined, the technique utilised, and the applications applied can all be used to classify data mining systems [7, 8].

1.2.2   Nature of Data

Data can be collected from multiple heterogonous repositories along with different file formats. They can be gathered from raw data, flat files, relational databases (RDBMS), transactional databases, the World Wide Web, data warehouses, etc. [9]. The database system may include spatial databases and so on. A different repository has different challenges while mining the data [10].

1.2.3   Data Mining Techniques

Data mining is a subset of AI and a superset of ML. Data mining techniques broadly classified into supervised learning and unsupervised learning. These two categories are broadly described in the following chapters. Data mining techniques or func-tionalities are defined to find interesting hidden patterns in data mining tasks. However, the data mining uses different abstraction levels to search patterns at vari-ous granularities [11–13]. The various data mining functionalities are dis-cussed below.

1.2.3.1 Classification Analysis

Classification is a process of data mining that is based on the analysis of the training data (i.e., objects whose class label is predefined). It can discover a model that can predict a class object with an unknown class label with the help of an expert set of historical class objects with known class labels. These classification models can be categorized into various forms, such as a k- nearest neighbor, neural network, Bayesian model, decision tree, random forest, support vector machine, etc. [16]. A kNN technique helps to predict the class object with the unknown class label by analyzing the most common class among k closest examples. It is a simple, intuitive method and provides a good classification result from any distribution. Naïve Bayes classification is simply based on the Bayesian probability theory. It is one of the fastest classification algorithms, which can handle real-life discrete data efficiently [14, 15]. A decision tree model performs classification using "IF-THEN" rules where a special tree data structure is used to apply that rule on each attribute value, each edge represents the result of the test and each leave represents a target class. Random forest is nothing but a collection of multiple decision trees to extract the benefits of all the efficient and effective decision tree sub-models. A neural network (NN) can also be used to perform classification. All these classification models are discussed in Chap. 3 extensively [17].

1.2.3.2 Case Study Example

Suppose you are a manager of a mobile phone shop. You aim is to discover some interesting buying patterns of customers and find some strategic ideas for increasing the growth of the purchase in the business by analyzing the previous purchasing or selling data and customer's buying patterns. If we apply the "IF-THEN" rules, we can divide the customers into some classes (excellent, good, poor). However, IF the age of a customer is "young" (20–30 years) and income is "high" along with a "good" CIBIL score, then he or she has an "excellent" probability to buy a mobile phone or belongs to the "excellent" class. IF the age of a customer is "young" (20–30 years) and income is "low" along with a "good" CIBIL score, then he or she has a "good" probability to buy a mobile phone or belongs to the "good" class. IF the age of a customer is "young" (20–30 years) and income is "low" along with a "bad" CIBIL score, then he/ she has a "poor" probability to buy a mobile phone or belongs to the "poor" class [18, 19]. IF the age of a customer is "middle-aged" (31–45 years) and income is "high" along with a "good" CIBIL score, then he/she has also an "excellent" probability of buying a mobile phone or belongs to the "excellent" class. Similarly, a series of IF-THEN classification rules can be explored for this problem. Figure 1.3 shows a classification model that can be represented by a decision tree, and Fig. 1.4 projects a classification model that can be represented by a neural network.

Fig. 1.3 Representation of
a classification model by a
decision tree

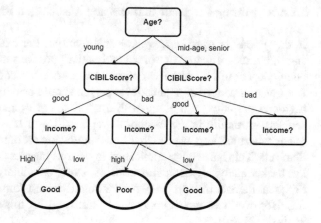

Fig. 1.4 Representation of
a classification model by a
neural network

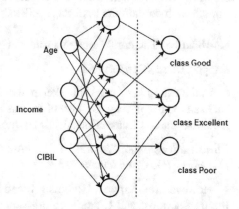

Example:

age (customer, "young") AND income(customer "High") AND CIBILscore (customer, "Good") => class (customer, "Excellent")

age (customer, "young") AND income(customer "Low") AND CIBILscore (customer, "Good") => class (customer, "Good")

age(customer, "young") AND income(customer "Low") AND CIBILscore(customer, "Bad") => class (customer, "Poor")

age(customer, "young") AND income(customer "High") AND CIBILscore(customer, "Bad") => class (customer, "Good")

age(customer, "midage/senior") AND income(customer "High") AND CIBILscore(customer, "Good") => class (customer, "Excellent")

age(customer, "midage/senior") AND income(customer "High") AND CIBILscore(customer, "Bad") => class (customer, "Good")

1.2.3.3   Mining Frequent Patterns and Association Rule Learning

In day-to-day transactions, we are very familiar. For example, "computer system and antivirus software" or "bread and butter" are very common patterns that frequently occur in a computer shop's or food court's transaction datasets, respectively. A sequential pattern of a game lover person should be a high-end PC followed by a high-end graphics card along with a good sound system. We can represent this sequence structure in the form of a graph, lattice or tree data structure. After frequent pattern mining, we can find some interesting set of rules associated with those item sets. This has a huge application in various business applications such as market basket analysis, medical treatments, stock data analysis, weblog mining, etc. Frequent itemset mining is the most straightforward form of frequent pattern mining. However, two metrics play the central roles for this kind of frequent itemsets analysis [21, 20].

Support   Support between two item sets A and B represents the number of transactions, which contain the item set A + B/total number of transactions.

Confidence   The confidence of a rule A≥B can be defined as

Support (A≥B)/Support (A).

In the example of market basket analysis, butter and bread are the frequent item set used to purchase by the most of the customers. An association rule can be mined from that market transactions dataset, which is

Purchase (customer, "bread") ≥ Purchase (customer, "butter")
[Support=1%, Confidence=50%]

However, the Support 1% means bread and butter purchased together for all of the transactions, and Confidence 50% means that if a customer purchases bread, then there is a 50% chance that he/she will buy butter along with it [21, 22]. It falls under the category of single-dimensional association rule as it adopts a single attribute like "purchase". We can mine an association rule in a computer shop as

Age ("customer", "25-35") & Income ("customer", "25K-40K") ≥ Purchase
("customer", "Laptop")

[Support=3%, Confidence=70%]

and it is called a multi-dimensional association rule. All association rules must support the minimum threshold values of support and confidence; otherwise, we should discard that rule. The association rule comes in helpful when it comes to analyzing datasets. Barcode scanners are used to capture data in supermarkets. These databases store a large number of transaction records [23, 24]. As a result, the manager will be able to establish whether specific sets of items are frequently purchased together and use this information to modify shop layouts, cross-sell, and promotions based on statistics and future business decisions. Algorithms suitable for finding frequent itemsets and association rules are broadly classified into two types (Fig. 1.5).

Fig. 1.5 Types of frequent
itemset mining

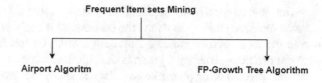

Fig. 1.6 A 2D plot of
arbitrary oval-shaped
clusters showing three
clusters along with "+"
cluster centres and outliers

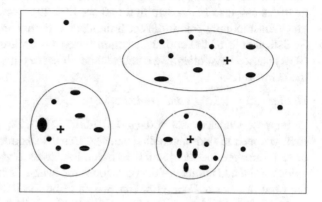

1.2.3.4   Clustering Analysis

Clustering is a formClustering analysis is a learning process through observations
rather than through examples. That's why it is called unsupervised learning, i.e., it
has a self-expertise to train or learn the data without class labels and can generate
such labels. It does not require the involvement of an expert during the learning
process. The real-life data may be labeled or unlabeled or a mix of both (semi-
labeled). However, clustering can be defined as the homogenous grouping of objects
rather than discarding other objects. Each cluster contains objects of similar kinds,
whereas the characteristics of objects of different clusters are entirely different. The
key requirement of a clustering process that it needs is a good measure of similarity
between data objects [26, 27]. Various clustering algorithms exist in the literature,
such as k-means, density-based, hierarchical, graph-based clustering, etc. Figure 1.6
shows a 2D plot of arbitrary oval-shaped clusters showing three clusters along with
"+" cluster centers and outliers.

The detail about the clustering process and unsupervised learning is described in
Chap. 4.

1.2.4   Anomaly or Outlier Detection

Figure 1.4 shows the outlier objects that reside outside of all the clusters. They
should not be part of any clusters as their behaviour is different from the behaviour
of the objects that reside within the clusters. These outliers treat as noisy objects and
need to be discarded. The outliers can be detected using statistical tests or

deviation-based methods or using distance measures. Figure 1.4 follows the distance measure strategy to detect the outliers. The analysis process of outlier data is referred to as outlier mining. This outlier analysis has huge applications in real-world problems. Making a graph is the simplest technique to find outliers. Outliers can be detected using plots like box plots, scatter plots, and histograms. Outliers can also be identified using the mean and standard deviation. The interquartile range and quartiles can also be used to find outliers [28, 29]. For example, outlier analysis plays an important role to detect fraudulent activities in credit card transactions. Statistical rule of the outlier detection based on a mean and interquartile range (IQR) is used most often. We can detect an object or entry as an outlier if it follows the below rules:

Outlier <M_{Q_1}-1.5xIQR and Outlier >M_{Q_2}+1.5XIQR

Suppose we have a set of data: 0.5, 20, 97, 21, 3, 28, 99, 30, 23, 32, and 24. At first, we need to find the median value in this distribution. To find the median, we have to arrange the distribution in ascending order and then choose the median value for the odd number of data or choose the average of the two middle values for the even number of data. However, we can find the interquartile range (IQR) by subtracting the median value of the 2nd half (Q_2) from the median value of the 1st half (Q_1), therefore, we have: IQR = M_{Q_2} − M_{Q_1}.

Data in ascending order: 0.5, 3, 20, 21, 23, 24, 28, 30, 32, 97, 99
Median value = 24
So Q_1 = 0.5, 3, 20, 21, 23 and Q_2 = 28, 30, 32, 97, 99
Now, M_{Q_1} = 20, M_{Q_2} = 32 and IQR = 32−20 = 12

Outlier < 18 − 1.5 × 12 = 1 and Outlier > 32 + 1.5 × 12 = 50.
So, in this given distribution 0.5 (<1) and 97 and 99 (>50) will be treated as outliers.

1.3   Regression Analysis and Prediction

Regression analysis is a set of statistical techniques for evaluating the connection between variables in statistical analysis. It's commonly utilized for forecasting and prediction. Regression analysis is also used to figure out which independent variables are related to the dependent variables and to investigate the types of correlations. The main difference between classification and prediction is that classification predicts discrete labels, whereas prediction creates models on continuous-valued function [30]. That means it is preferred to find missing numeric values rather than class labels. Regression analysis is a statistical procedure that is mainly used for numeric prediction. There are mostly two types of regression analysis, (i) linear regression and (ii) non-linear regression.

1.3.1   Sequence Analysis

Time series data mining leads to sequential pattern analysis. It describes and predicts trends in the behavior of items that vary over time. This includes association analysis, correlation, classification, prediction and clustering of time-related objects. For example, stock market data of the last several years are time-series data. This kind of sequential pattern mining helps to identify stock evolution regularities for overall stocks and predict future trends on market prices so that you can make some intelligent decisions regarding stock investments [31].

1.3.2   Limitations of Data Mining

Data mining has a lot of advantages when we can use it in specific industries such as marketing, governments, finance, manufacturing, etc. Besides that, it has several disadvantages related to privacy, security, etc. The scope of this book addresses various issues in data mining methodology.

While data mining software is helpful, it is not a stand-alone application. Data mining requires highly skilled technical and analytical professionals who can arrange the analysis and interpret the results to be successful. As a result, the limitations of data mining are mostly data or personnel-related rather than technological. While data mining can aid in the finding of patterns and connections, it does not provide information to the user regarding the value or importance of these patterns. The user must make these kinds of selections.

Data is valuable until it is secured. Data mining generally deals with a massive volume of data to discover some interesting and valuable information that can increase the growth of the business. However, how much the information of employees and customers in the business is taken care of is still in question. Hackers can steal that useful financial information at any time [32, 33].

Information used in data mining can not only be stolen but can be misused or exploited by unethical people or businesses to take the benefit of vulnerable people. Data collection at the initial stage of the data mining process can be overwhelming with some additional irrelevant or noisy data. This noisy data can confuse the whole process and sometimes force the model to overfit the data. As a result, it may create a poor impact on the accuracy of the model. Therefore, it is a great challenge to handle that unnecessary information while applying data mining techniques. Trees, graphs, charts, rules, tables and other interactive and expressive knowledge representation approaches are required by data mining systems. To be effective, data mining algorithms must be scalable. A dataset contains a large volume of data, the execution time of data mining algorithms must be predictable and acceptable. These are the most important considerations while creating together a data mining system [34].

1.4   Applications of Data Mining

Although data mining software is helpful, it is not a stand-alone program. Data mining requires highly qualified technical and analytical expertise to arrange the analysis and understand the results to be successful. As a result, the limitations of data mining are usually data or personnel-related rather than technological. Although data mining can aid in the finding of patterns and connections, it does not offer the user information regarding the value or importance of these patterns. The user must make such decisions [35–37]. "Figure 1.7" describes the different fields through which data mining can spread its applications.

Data mining Large consumer-based firms mostly use data mining in the retail, financial, communication and marketing areas. It is used to determine price, client preferences, and product placement, impacting sales, customer satisfaction, and company profitability. The following is a list of the most important applications of data mining:

Personal Grooming and Healthcare
In the realm of healthcare, data mining has a significant impact. It employs data and analytics to determine the most effective techniques for improving treatment and lowering costs. To make things easier for patients, scientists use multidimensional databases, machine learning, soft computing, data visualization, statistics, and other data mining techniques [38, 39]. We can forecast the number of patients in each category using data mining and ensure that they receive the necessary care at the right place and at the right time.

Analysis of the Market Basket
This modeling technique is based on the idea that you're more likely to buy another set if you buy one set of things. Using this strategy, a retailer can learn about a customer's buying habits and adjust the store's layout to meet their demands [40].

Education and Skills
Educational data mining is used to identify and predict future learning behavior in children. If a student is enrolled in a specific course, the institutes can use data mining to determine which similar courses they would be interested in later. It's also

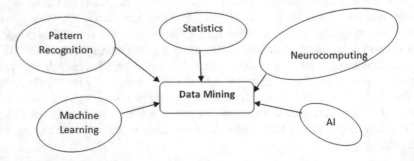

Fig. 1.7 Data mining multidisciplinary

helpful to concentrate on what to teach and how to teach. Students' learning patterns can be captured by institutes, which can then be used to build teaching strategies for them.

Engineering for Manufacturing
We can identify patterns in complex industrial processes by employing data mining technologies. We may use this to anticipate the duration, cost and dependencies of product development, among other things.

Detection of Fraud
Data mining We can use data mining techniques to protect all users' information by creating a flawless fraud detection system. We can classify fraudulent and non-fraudulent data and create an algorithm to determine whether a record is fraudulent or not using data mining.

Customer Relationship Management
We can use data mining techniques to maintain a proper relationship with a customer.
Some other areas where data mining is used:

- Intrusion detection
- Lie detection
- Customer segmentation
- Financial banking
- Corporate surveillance
- Research analysis
- Criminal investigation
- Bioinformatics

1.5 Incremental Data Mining

The term "incremental" refers to modifications to an existing database, such as adding new data or deleting existing data. This process is also known as the "percentage of delta change in the database". The critical topic in today's world, because the vast majority of databases today are dynamic. As a result, we'll need to create some new data mining approaches (algorithms) that can deal with the database's dynamic nature efficiently and effectively. The goal of incremental data mining techniques is to reduce the time it takes to scan and calculate new records. The efficiency of the newly introduced record update problem is improved here. These are the main elements that lead to the usage of incremental mining:

(i) The database is constantly updated.
(ii) The database must be scanned again after each modification (insertion and deletion). As a result, rescanning takes a long time.
(iii) As a result, the incremental version of the clustering method comprises separate dynamic operations for insertion and deletion.

(iv) Databases may have regular changes and hence be dynamic at any time.
 (v) The existing clustering method must be modified after database insertions and deletions.

    After that, we believed that the incremental approach method would perform better than the actual one.

1.5.1 Benefits of Incremental Data Mining

The incremental data mining concept is innovative and fruitful for today's busy life. It provides several advantages, such as:

  (i) The incremental approach helps us to efficiently handle many changes (insertion or deletion of data) in the databases.
 (ii) In the dynamic environment, the database always gets modified, and during this modification, the database needs to be re-clustered. Thus, this process requires a lot of time for clustering. The incremental data mining approach provides a fast and easy way to re-cluster the database. Thus, it saves time.
(iii) It not only saves time but also saves effort by using fewer computations.
(iv) With the help of incremental data mining, we can verify which part of an existing clustering is affected by an update of the database. We can also examine the behavior of the cluster metadata (centres, radius, number of objects, etc.) during this updation in the database.

1.6 Regression Methods

When a model is built for a continuous-valued function, or ordered value, rather than a categorical label, it is referred to as a predictor or regressor. Regression analysis is a statistical technique that is commonly used to make numerical predictions. For example, regression analysis is commonly employed in stock market growth and risk forecasting, weather forecasting and temperature forecasting. It describes the relationship between one or more numeric independent variables $(x_1, x_2, \ldots, x_n)$ and a single dependent variable (y). This relationship is now assumed to be a straight line. A line is represented by the equation $y = mx + b$ in algebra, where m is the slope, which specifies how much the line rises for each increase in x. When x becomes 0, b determines the value of y:

$$m = r \cdot \frac{S_y}{S_x} \cdot r = \frac{1}{n-1} \Sigma \left(\frac{x_i - x'}{S_x} \right) \left(\frac{y_i - y'}{S_y} \right)$$

where, r is a correlation coefficient, x' and y' are sample means and S_x, S_y are sample standard deviations.

1.6.1 Illustration

Given, $(x, y) = \{(1,1), (2,2), (2,3), (3,6)\}$ are set of observations in a two-dimensional feature space. Our aim is to calculate the final regression equation in the form of a straight line:

$$y = mx + b$$

After calculating the required parameters for m, $x' = 2$, $S_x = 0.816$, $y' = 3$ and $S_y = 2.160$, we have:

$$r = \frac{1}{n-1} \sum \left(\frac{x_i - x'}{S_x} \right) \left(\frac{y_i - y'}{S_y} \right)$$

$$\Rightarrow r = 0.946$$

$$m = 0.946 * \frac{2.160}{0.816}$$

$$\Rightarrow m = 2.50$$

Therefore, our next objective is to calculate the value of b. We calculate it as follows:

$$y' = mx + b$$

$$\Rightarrow 3 = 2.50 * 2 + b$$

$$\Rightarrow b = -2$$

So, the final regression line is $y' = 2.50*x - 2$; for any value of new incoming data x, we will find an outcome y'. This overall technique is called simple linear regression. The estimation method ordinary least squares (OLS) is used to find the best estimations of m and b. In this approach, m and b are chosen to minimize the sum of squared errors (SSE), or the vertical distance between the anticipated and actual y values. The term "residuals" refers to these faults. Figure 1.8 shows a typical linear regression example with residuals.

The error term can be evaluated as

$$\text{error} = \sum_{i=1}^{n} \left(y_i - y_i' \right)^2$$

The accuracy of a linear regression model can be checked with the help of residual standard error (RSE) parameter. A model is said to be a good fit if RSE is small:

$$\text{RSE} = \sqrt{\frac{1}{n-2} * \sum_{i=1}^{n} (y_i - y')^2}$$

Fig. 1.8 A typical linear regression example with residuals

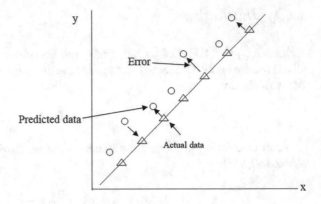

A multiple linear regression can be expressed as

$$y = W_0 + W_1X_1 + W_2X_2 + \varepsilon_i$$

where, ε_i is the error term. A polynomial non-linear regression can be expressed as

$$y = W_0 + W_1X^1 + W_2X^2 + W_3X^3 + \cdots\cdots$$

So, we convert it into linear form using the transformations, $X^2 = X_1$, $X^3 = X_2$, and so on.

$$y = W_0 + W_1X_1 + W_2X_2 + \varepsilon_i$$

to import the dataset and load it into our pandas dataframe. The Naïve Bayes, decision tree (CART), SVM, random forest, ensemble voting, AdaBoost and XGBoost algorithms are sufficient to create an instance of *KNeighborsClassifier*, *GaussianNB*, DecisionTreeClassifier, SVC, RandomForest, *VotingClassifier*, *AdaBoostClassifier* and *XGBClassifier*, respectively.

dataset1=pd.read_csv("featrain.csv");
print(dataset1)
X = dataset1.iloc[:,0:28].values
Y = dataset1.iloc[:,28].values

dataset2=pd.read_csv("featest.csv");
print(dataset2)
X_validation = dataset2.iloc[:,0:28].values
Y_validation= dataset2.iloc[:,28].values
The accuracy is KNN: 0.526195
The accuracy is NB: 0.622357
The accuracy is CART: 0.545960
The accuracy is XG: 0.561987
The accuracy is Ada: 0.538418

Now, we have applied the popular classifiers KNN, Naive Bayes and SVM on the testing data to check the accuracy score, confusion matrix, and classification report, respectively. Among all these classifiers, random forest (RF) and Naïve Bayes (NB) are the two best classification accuracies on the validation dataset. Their accuracy is 68.03% and 67.21%, respectively. We have applied all the classifiers one by one for testing purposes. The above program snippet shows the execution for almost all popular classifications and the classification report. This section has applied ensemble voting classifier and boosting techniques on our experimented datasets.

1.6.2 Assumptions of Linear Regression

A linear relationship must exist between y and x. There must be no or very little multi-collinearity value between independent variables. Homoscedasticity states that the errors are equally distributed, whereas heteroscedasticity states that the errors are not equally distributed. To correct this phenomenon, a log function is used. Besides these linear and non-linear regressions, some other kinds of regression exist, such as logistic polynomial regression, Poisson regression, logistic regression, ridge regression, lasso regression, etc., some brief discussions about some necessary terms in regression.

Regularisation
It helps to solve overfitting problems in machine learning. A simple model would be inferior if it overfitted with low accuracy. Regularisation shrinks the coefficients estimates towards zero.

Ridge Regression
We will describe a small amount of bias to show that a new line will fit the data. In Fig. 1.9, the linear regression is not a suitable fit as the sum of squares for the training data is zero, but RSS for the testing data is large. In this situation, a ridge regression can provide better and long-term predictions with low variance by starting with a slightly worst fit. It has some important features:

- Ridge regression's line end up with smaller slope than linear regression.
- Ridge regression minimises RSS and ($\lambda \times slop^2$), where λ is a penalty term, and, typically, tenfold cross validation is used to determine the value of λ.
- If there are 1000 features but only 500 observations, then it is better to use ridge regression.

Lasso Regression
It is a type of linear regression but with shrinkage flexibility. It is suitable for sparse and straight forward datasets, and it can shrink the data values towards the mean. Ridge regression can only reduce the slope asymptotically near zero, whereas lasso regression can ultimately reduce the slope to zero. Therefore, instead of squaring

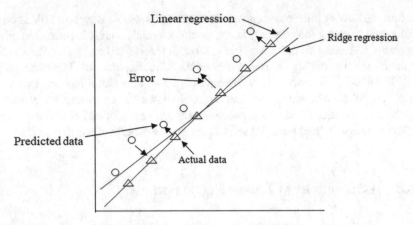

Fig. 1.9 A comparison between linear regression and ridge regression

Fig. 1.10 A typical
S-shaped curve for logistic
regression where number
in between $\{-5,+5\}$

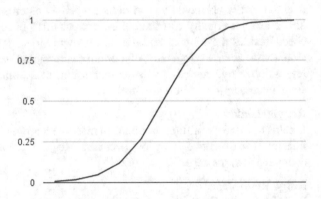

the $(\lambda \times \text{slop}^2)$, we take $(\lambda \times |\text{slop}|)$ in lasso regression and $\lambda \epsilon \{0,+\text{ve}\}$. It also provides less variance.

Logistic Regression
For discovering binary categorical outcomes, logistic regression is utilized. It is based on the sigmoid function notion. Figure 1.10 shows a typical S-shaped curve for logistic regression.

It can be expressed as

$$y = \frac{1}{1+e^{\text{value}}}$$

$$\frac{P(x)}{1-P(x)} = e^{b_0+b_1*x}$$

$$y = \frac{e^{b_0+b_1*x}}{1+e^{b_0+b_1*x}}$$

where y is the predicted output, x is the input variable, b_0 is the bias and b_1 is the coefficient for the input value x. We can compute the values of b_0 and b_1 with the help of maximum likelihood function. Finally, we can turn around the above equation as follows:

$$b_0 + b_1 * x = \log\left(\frac{P(x)}{1 - P(x)}\right)$$

This is also called as logit or log-odds representation.

1.7   Evaluating Model Performances

The easiest way to evaluate a classifier's performance is to see if it accomplishes its goals. As a result, it is critical to establish model performance metrics that focus on utility rather than absolute accuracy. The confusion matrix is one such metric. A confusion matrix is a table that classifies predictions based on whether or not they match the data's actual value. As a result, one dimension of the table represents expected values, while the other dimension represents actual values. The most common way to describe the binary classification model is with a 2X2 confusion matrix. The classification is correct when the projected value matches the actual value in this confusion matrix. The correct predictions appear on the diagonal of the confusion matrix, designated by TP and TN. In contrast, the off-diagonal cells of the confusion matrix represent the disparity between predicted and actual values, resulting in incorrect categorization. The model's ability to distinguish one class from others is one of the most popular performance measures. The positive class is known as the class of interest, whereas the negative class is known as everything else. The following confusion matrix depicts the link between positive and negative class predictions.

ROC Curve
The receiver operating characteristic curve is a popular tool for examining the trade-off between detecting true positives and minimizing false positives. Engineers in the communication field developed ROC curves during World War II because the receivers of radar and radio transmissions needed a way to distinguish between actual signals and false alarms. Today, the same technique may be used to visualize the efficacy of machine learning models. It is worthy to note that the ROC curves are appropriate when the number of observations in each class is evenly distributed. When the number of observations in each class is evenly distributed, It is worthy to note that the ROC curves are appropriate when the number of observations in each class is evenly distributed. Figure 1.11 shows a typical example of a ROC curve plot.

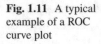

Fig. 1.11 A typical example of a ROC curve plot

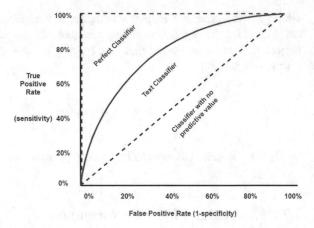

The x-axis represents the proportion of false positives (1-specificity), and the y-axis represents the proportion of genuine positives (sensitivity) in this ROC curve. Three potential classifiers will be considered here. The diagonal dotted line in this diagram depicts a classifier with no predictive value. This classifier finds genuine positives and false positives that are identical. This line serves as a benchmark against which the other classifiers are measured. Any ROC curve that falls near this line suggests that the models are ineffective. The ideal classifier has a curve with a true positive rate of 100 percent. In practice, most of the real-world classifiers are comparable to the test classifier, which is midway between ideal and worthless. The better the curve is at identifying positive values, the closer it is to the perfect classifier. The area under the ROC curve is a statistical metric (AUC). AUC ranges from 0.5 (for a classifier with no predictive value) to 1.0 (for a classifier having a predictive value) (for an ideal classifier).

Bias-Variance Trade-off

Bias is an error caused by the learning algorithm's incorrect or excessively simplified assumptions. This can cause the model to underfit your data, making it difficult for it to forecast accurately and for you to transfer your knowledge from the training set to the test set. Variance is an error caused by the learning algorithm's overcomplication. As a result, the algorithm becomes extremely sensitive to large amounts of variation in your training data, perhaps causing your model to overfit the data. The bias-variance decomposition adds the bias, variance and a bit of irreducible error due to noise in the underlying dataset to deconstruct the learning error from any algorithm [25]. Essentially, suppose we make the model more complex and include more variables. In that case, we will lose bias but acquire some variance – we will have to trade off bias and variance to get the ideally decreased error level. We don't want your model to have a lot of bias or variance.

1.7.1 Conclusion

The main issues lie in improving the classification or prediction process's accuracy, efficiency, and scalability. That's why extensive data preprocessing is required. Data in the real world is dirty. Many potentially incorrect data may exist, such as noise or errors, containing discrepancies in names, disguised missing data, etc. We can handle the missing data problem by ignoring the tuple when the class label is missing. Besides that, we can also fill in the missing values by global constants or attribute mean or the most probable values with the help of the Bayes' theorem. We can handle the noisy data problem using Binning, Regression, or clustering techniques. An input dataset may contain irrelevant attributes that can create redundancy. Correlation analysis can be used to identify the statistically strong relevance between two attributes, and any one attribute could be removed from further analysis. Data reduction creates a smaller version of the data collection that gives the same (or nearly the same) analytical results. Data can also be altered by applying higher-level concepts to it. It can be accomplished through the use of concept hierarchies. This is especially beneficial for properties that have a constant value. Because generalisation compresses the original training data, learning may require fewer input/output procedures. Many other approaches, such as wavelet transformation and principal components analysis (PCA) and discretization techniques like binning, histogram analysis, and clustering, can be used to decrease data.

1.8 Exercise

Q1. What is Data mining? Explain how the evolution of database technology influences data mining.

Q2. Under the process of KDD, how data mining steps can be described?

Q3. Suppose your task as a super of a hospital is to design a data mining system to examine the hospital medicine stock databases and staff databases where it contains necessary information of each staff and some life-saving drugs. Describe the architecture you would choose for each database.

Q4. "Incremental data mining can increase the power of handling the analysis process of dynamic real-life databases" – Justify this statement.

Q5. What are the various kinds of data on which data mining techniques are applicable?

Q6. How data warehouse is dependent on the data mining systems to help the business executives to make strategic decisions?

Q7. Define the various benefits of using the concept of incremental data mining.

Q8. Describe the various application domains of data mining. Explain any two-application domains in detail where data mining techniques play the key roles.

1.9   Interview Questions

Q1. How business intelligence is related to data mining?

Q2. Explain the different stages of data mining.

Q3. What do you understand by discrete and continuous data in data mining?

Q4. What are the most significant disadvantages of data mining?

Q5. What do you understand by DMX in the context of data mining?

Q6. What are the foundations of data mining?

Q7. Name the steps used in data mining?

Q8. What are required technological drivers in data mining?

Q9. Give an introduction to data mining query language?

Q10. What do you mean by preprocessing of data in data mining?

Q11. What are the different fields where data mining is used?

Q12. What do you understand by data purging?

Q13. What are the different problems that "data mining" can solve?

Q14. How do data mining and data warehousing work together?

Q15. What do you understand by the time series algorithm in data mining?

References

1. Eshref Januzaj, Hans-Peter Kriegel, Martin Pfeifle, "Towards Effective and Efficient Distributed Clustering", Workshop on Clustering Large Data Sets (ICDM2003), Melbourne, FL, 2003.
2. S.Jiang, X.Song, "A clustering based method for unsupervised intrusion detections" . Pattern Recognition Letters, PP.802-810, 2006.
3. Guha A., D. Samanta, A. Banerjee and D. Agarwal, "A Deep Learning Model for Information Loss Prevention From Multi-Page Digital Documents," in IEEE Access, vol. 9, pp. 80451–80465, 2021, doi: https://doi.org/10.1109/ACCESS.2021.3084841.
4. A.M.Sowjanya, M.Shashi, "Cluster Feature-Based Incremental Clustering Approach (CFICA) For Numerical Data, IJCSNS International Journal of Computer Science and Network Security, VOL.10 No.9, September 2010.
5. Air-pollution database, WBPCB, URL: 'http://www.wbpcb.gov.in/html/airqualitynxt.php'.
6. Althar, R.R., Samanta, D. The realist approach for evaluation of computational intelligence in software engineering. Innovations Syst Softw Eng 17, 17–27 (2021). https://doi.org/10.1007/s11334-020-00383-2.
7. Anil Kumar Tiwari, Lokesh Kumar Sharma, G. Rama Krishna, " Entropy Weighting Genetic k-Means Algorithm for Subspace Clustering ",International Journal of Computer Applications (0975– 8887),Volume 7– No.7, October 2010.
8. Aristidis Likasa , Nikos Vlassis, Jakob J. Verbeek ," The global k-means clustering algorithm " , the journal of the pattern recognition society, Pattern Recognition36 (2003) 451–461, 2002.
9. B. Naik, M. S. Obaidat, J. Nayak, D. Pelusi, P. Vijayakumar and S. H. Islam, "Intelligent Secure Ecosystem Based on Metaheuristic and Functional Link Neural Network for Edge of Things," in IEEE Transactions on Industrial Informatics, vol. 16, no. 3, pp. 1947–1956, March 2020, doi: https://doi.org/10.1109/TII.2019.2920831.
10. Carlos Ordonez and Edward Omiecinski, "Efficient Disk-Based K-Means Clustering for Relational Databases", IEEE transaction on knowledge and Data Engineering,Vol.16,No.8,August 2004.

11. Carlos Ordonez, "Clustering Binary Data Streams with K-means", San Diego, CA, USA. Copyright 2003, ACM 1- 58113-763-x, DMKD'03, June 13, 2003.
12. CHEN Ning , CHEN An, ZHOU Long-xiang, "An Incremental Grid Density-Based Clustering Algorithm", Journal of Software, Vol.13, No.1,2002.
13. D. Samanta et al., "Cipher Block Chaining Support Vector Machine for Secured Decentralized Cloud Enabled Intelligent IoT Architecture," in IEEE Access, vol. 9, pp. 98013–98025, 2021, doi: https://doi.org/10.1109/ACCESS.2021.3095297.
14. Data Mining concepts and techniques by Jiawei Han and Micheline Kamber, Morgan Kaufmann (publisher) from chapter-7 'cluster analysis', ISBN:978-1-55860-901-3, 2006.
15. Debashis Das Chakladar and Sanjay Chakraborty, EEG Based Emotion Classification using Correlation Based Subset Selection, Biologically Inspired Cognitive Architectures (Cognitive Systems Research), Elsevier, 2018.
16. Dunham, M.H., Data Mining: Introductory And Advanced Topics, New Jersey: Prentice Hall, ISBN-13: 9780130888921. 2003.
17. Govender, P., & Sivakumar, V. (2020). Application of k-means and hierarchical clustering techniques for analysis of air pollution: A review (1980–2019). Atmospheric pollution research, 11(1), 40–56.
18. Guha, A., Samanta, D. Hybrid Approach to Document Anomaly Detection: An Application to Facilitate RPA in Title Insurance. Int. J. Autom. Comput. 18, 55–72 (2021). https://doi.org/10.1007/s11633-020-1247-y
19. H.Witten, Data mining: practical machine learning tools and techniques with Java implementations San-Francisco, California : Morgan Kaufmann,ISBN: 978-0-12-374856-0 2000.
20. Jahwar, A. F., & Abdulazeez, A. M. (2020). Meta-heuristic algorithms for k-means clustering: A review. PalArch's Journal of Archaeology of Egypt/Egyptology, 17(7), 12002–12020.
21. K. Mumtaz, Dr. K. Duraiswamy, "An Analysis on Density Based Clustering of Multi Dimensional Spatial Data", Indian Journal of Computer Science and Engineering, Vol. 1 No 1, pp-8–12, ISSN : 0976-5166.
22. K. Wang et al., "A Trusted Consensus Scheme for Collaborative Learning in the Edge AI Computing Domain," in IEEE Network, vol. 35, no. 1, pp. 204-210, January/February 2021, doi: https://doi.org/10.1109/MNET.011.2000249.
23. Kantardzic, M.Data Mining: concepts, models, method, and algorithms, New Jersey: IEEE press, ISBN: 978-0-471-22852-3, 2003.
24. Kehar Singh, Dimple Malik and Naveen Sharma, "Evolving limitations in K-means algorithm in data Mining and their removal", IJCEM International Journal of Computational Engineering & Management, Vol. 12, April 2011.
25. Khamparia, A, Singh, PK, Rani, P, Samanta, D, Khanna, A, Bhushan, B. An internet of health things-driven deep learning framework for detection and classification of skin cancer using transfer learning. Trans Emerging Tel Tech. 2020;e3963. https://doi.org/10.1002/ett.3963
26. Long, Z. Z., Xu, G., Du, J., Zhu, H., Yan, T., & Yu, Y. F. (2021). Flexible Subspace Clustering: A Joint Feature Selection and K-Means Clustering Framework. Big Data Research, 23, 100170.
27. Lopamudra Dey, Sanjay Chakraborty, Anirban Mukhopadhyay. Machine Learning Techniques for Sequence-based Prediction of Viral-Host Interactions between SARS-CoV-2 and Human Proteins. Biomedical Journal, Elsevier, 2020.
28. Martin Ester, Hans-Peter Kriegel, Jorg Sander, Michael Wimmer, Xiaowei Xu, "Incremental clustering for mining in a data ware housing", 24th VLDB Conference New York, USA, 1998.
29. Michael K. Ng, Mark Junjie Li, Joshua Zhexue Huang, and Zengyou He, " On the Impact of Dissimilarity Measure in k-Modes Clustering Algorithm ", IEEE transaction on pattern analysis and machine intelligence, vol.29, No. 3, March 2007.
30. Naresh kumar Nagwani and Ashok Bhansali, "An Object Oriented Email Clustering Model Using Weighted Similarities between Emails Attributes", International Journal of Research and Reviews in Computer science (IJRRCS), Vol. 1, No. 2, June 2010.

31. Oyelade, O.J, Oladipupo, O. O, Obagbuwa, I. C, "Application of k-means Clustering algorithm for prediction of Students' Academic Performance",(IJCSIS) International Journal of Computer Science and Information security,Vol.7,No. 1, 2010.
32. Rohan Kumar, Rajat Kumar, Pinki Kumar, Vishal Kumar, Sanjay Chakraborty, Prediction of Protein-Protein interaction as Carcinogenic using Deep Learning Techniques, 2nd International Conference on Intelligent Computing, Information and Control Systems (ICICCS), Springer, pp. 461–475, 2021.
33. Sauravjyoti Sarmah, Dhruba K. Bhattacharyya," An Effective Technique for Clustering Incremental Gene Expression data" , IJCSI International Journal of Computer Science Issues, Vol. 7, Issue 3, No 3, May 2010.
34. Steven Young, Itemer Arel, Thomas P. Karnowski,Derek Rose, University of Tennessee, "A Fast and Stable incremental clustering Algorithm", TN 37996, 7th International 2010.
35. Taoying Li and Yan Chen, "Fuzzy K-means Incremental Clustering Based on K-Center and Vector Quantization", Journal of computers, vol. 5, No.11, November 2010.
36. Tapas Kanungo , David M. Mount , "An Efficient k-Means Clustering Algorithm: Analysis and implementation IEEE transaction vol. 24 No. 7, July 2002.
37. Tavallali, P., Tavallali, P., & Singhal, M. (2021). K-means tree: an optimal clustering tree for unsupervised learning. The Journal of Supercomputing, 77(5), 5239–5266.
38. Weka, Waikato environment for knowledge environment – http://www.cs.waikato.ac.nz/ml/weka/.
39. Xiaoke Su, Yang Lan, Renxia Wan, and Yuming, "A Fast Incremental Clustering Algorithm ", international Symposium on Information Processing (ISIP'09), Huangshan, P.R.China, August-21–23,pp:175–178,2009.
40. Zuriana Abu Bakar, Mustafa Mat Deris and Arifah Che Alhadi, "Performance analysis of partitional and incremental clustering", SNATI, ISBN-979-756-061-6, 2005.

Chapter 2
A Brief Concept on Machine Learning

2.1 Introduction

An unbreakable bond between human beings and machines has existed since the last few eras. It could be thought of as science fiction stories in the early stages when computers seem to be a controller that can play easy tic-tac-toe and chess games. Later, traffic lights and communications are given to robots, followed by military drones and weapons. At that time, an obvious question came into mind: Can a machine think and behave like a human? Or can a computer build a strong knowledge base with some decision-making capabilities based on some previous experiences? We can say "YES" after enriching ourselves with the knowledge of "machine learning" [1, 2].

2.1.1 Case Study

"Before Machine Learning Human Learning Came"

Let's say a Professor teaching in an engineering college for the last couple of years. So, he used to teach the 3rd year undergraduate students for the previous four months. Now, his class consists of a total of twenty students. He knows the behaviour and attitude of each student, as they have been familiar with him for the last four months. Assume that a student "Deepak" wants to talk to his friend "Gita" during the lecture hours. Deepak exhibits some different kinds of behavior (shaking his legs or nail-biting, etc.), which he generally uses to avoid. His class Professor notices it before he starts gossiping with Gita. Suddenly, his class Professor warns him not to talk with Gita during his lecture time; Deepak is astonished and confused. Deepak is thinking about how his class Professor predicts even before it happens?

© The Author(s), under exclusive license to Springer Nature   23
Switzerland AG 2022
S. Chakraborty et al., *Data Classification and Incremental Clustering in Data Mining and Machine Learning*, EAI/Springer Innovations in Communication and Computing, https://doi.org/10.1007/978-3-030-93088-2_2

The Professor knows Deepak's behavior very well and stores it in his brain or knowledge base to collect past experiences. In addition, he can predict the outcome (Deepak's future movements) by analyzing those past experiences and behaviors. This exceptional capability of a human being to identify any future event based on past experiences is called "human learning" [3, 4]. However, it is possible for that Professor to remember and understand the behavior, characteristics, and attitudes of twenty students in his class, but if the Principal asked him to do a similar sort of learning for all two thousand students of the entire college, then the brain capacity of that Professor will fail to store such a huge volume of information. Then "machine learning" appears along with high computation power that will do the same job for any number of students and any amount of information [5]. In ML lingo, these past experiences are called previous data. Multiple definitions of ML can be given, such as the following.

2.1.2 Interesting Definitions

- "ML" is a topic of study that focuses on the creation of computer algorithms for converting data into intelligent actions.
- ML = data + computing power.
- ML is a branch of AI that enables computers to learn and improve without being explicitly programd.
- ML is a branch of AI that proposes that a computer program can adapt to new data without the intervention of a person.
- ML is a method of interpreting data, learning from it and then applying what they've learnt to create a judgement [6].

2.2 Artificial Intelligence (AI) vs. Data Mining vs. ML vs. Deep Learning

Data mining does not require ML methods, while the contrary is not true. In other words, you can use ML for jobs that do not require data mining, but you are virtually certainly utilizing ML if you are using data mining methods. The above four terms are strongly connected but not the same [7].

AI includes all these fields. AI is a branch of Computer Science where machines are programmed and can think and mimic actions like humans, and animals. With the help of AI, we can do speech and pattern recognition, computer vision, and problem-solving. ML is a type of AI in which a computer can make automatic decisions by analyzing, understanding, and identifying patterns in trained data. Nowadays, many big companies use ML to provide a better experience to their customers. For example, Uber uses ML techniques for familiar locations visit with cheap cost or better promotions. Similarly, Amazon uses ML techniques to give better product choice recommendations to their customers. Netflix uses ML techniques to provide better suggestions to their users regarding TV series and movies. Deep learning is a subset of ML. In deep learning, a deep multilayer perceptron concept is used to automatically rectify the inaccurate prediction during model learning by the model itself. Still, the programmer needs to fix that problem explicitly for ML-based approaches [8, 9].

2.2.1   An Illustration

Consider the following scenario to comprehend both ML and deep learning better. We might speculate that AI will be achieved in the future through ML and deep learning. Data mining and machine learning, on the other hand, are both subdomains of Data Science [10, 11]. As a result, they are inextricably linked. Data mining is a critical component of machine learning, and it is used to uncover significant patterns and trends hidden in massive amounts of data. Advanced algorithms are used in data mining and ML to find relevant data patterns. DM focuses on identifying unknown qualities in the data, while ML focuses on prediction based on known features learned from training data. DM is used to extract rules from existing data, whereas ML educates the computer to learn and comprehend the rules [12–14]. Despite the fact that data mining and ML overlap, they have a lot of variations in terms of how they're applied. The logical block diagram to represent the relationship among these terms has been shown in Fig. 2.1.

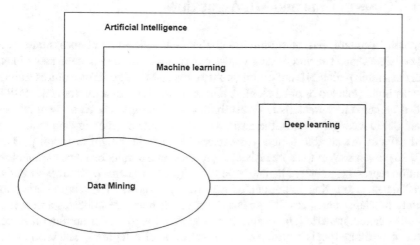

Fig. 2.1 The relationship among AI, ML, DL, and DM

2.3   How to use ML?

The following steps help to apply the ML algorithms to solve a problem.

1. **Collection of data**: You'll need to gather the data in an electronic format, like text file, spreadsheet, or SQL database. The data will be the raw material for an ML algorithm to generate actionable knowledge [14].
2. **Exploration and preparation of data:** Any ML result's quality is primarily determined by the data it uses. This stage of the ML process usually necessitates a significant amount of human intervention. Data accounts for 80% of ML effort, according to a widely quoted figure. During data exploration, a large portion of this time is spent learning more about the data and its subtleties [15].
3. **Developed a model**: By the time the data is ready for analysis, you'll probably have a good idea of what you want to learn from it. The specific ML goal will guide the choice of a suitable method, and the algorithm will use a model to represent the data.
4. **Model performance evaluation**: Each ML model produces a biased answer to the learning problem, and it's critical to assess how well the algorithm learned from its previous experiences. You may be able to evaluate the model's accuracy using a test dataset, or you may need to build performance metrics relevant to the intended application, depending on the type of model utilized [16].
5. **Improving model performance**: If improved performance is required, more advanced procedures must be used to supplement the model's performance. It may be essential to move to a different model altogether at times. As in step two of this method, you may need to supplement your data with more data or undertake further preliminary work [17, 18].

2.4   Types of Data and ML Algorithms

Choosing a suitable ML algorithm is a challenging endeavor that necessitates careful consideration. The sort of data you're examining, and the intended task at hand will determine which ML algorithm you use. Every ML algorithm requires training data as input (labeled or unlabeled). A dataset in the form of a matrix is accessible, and it contains a characteristic or attribute of an example called a "feature". For example, in a cancer dataset, the attributes could include genomic data from biopsied cells or measurable patient parameters like weight, height or blood pressure [19, 20]. Each row in the spreadsheet represents an example or instance, and each column represents a feature in matrix data. In Fig. 2.2, a snapshot of the well-known "Iris" dataset from Kaggle or the UCI ML library is shown. The features come in a variety of shapes and sizes. It's unsurprising that if a feature indicates a numerical trait, it's called numeric. The feature is considered categorical or nominal if it measures an attribute that is represented by a collection of categories. The term ordinal refers to a nominal variable having categories in an ordered list, a specific case of

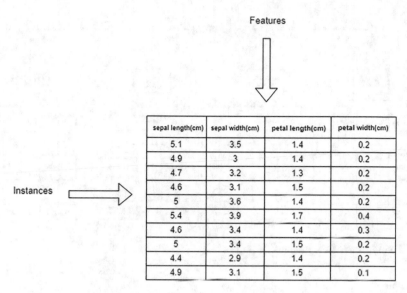

Fig. 2.2 Snapshot of an Iris dataset

categorical variables. Clothing sizes such as small, medium, and large and a rating of customer satisfaction scale of one to five are examples of ordinal variables. It's crucial to think about the meaning of features because the type and number of features in your dataset will help you choose the correct ML algorithm for the job. When the training set is tiny, a model with a correct bias and low variance performs better because it is less prone to overfitting. For example, the Naive Bayes classifier works well when the training set is huge. Low-bias, high-variance models perform better because they can handle complex relationships [21, 22].

When a human baby is born, he or she requires proper training from his or her parents to do day-to-day activities like walking, talking, eating, etc. When a spawn is born, it can learn to swim automatically; otherwise, it drowns. However, from this case study, we understand that a human baby requires a supervisor for its initial training, but a spawn does not require any supervisor for its initial survival. So, we consider the first case as a supervised learning process in nature and the second case as an unsupervised learning process in nature [23–25].

2.4.1 Supervised Learning

In a predictive model, a target class can be predicted with the help of a set of input features and efficient supervision. Past occurrences could be predicted using a predictive model. During rush hour, predictive models could be utilized to manage traffic signals in real-time. So, supervised learning, or more specifically classification and regression, is the process of developing a prediction model [26]. For instance, we can use a predictive model (classifier) to predict if a credit card user is

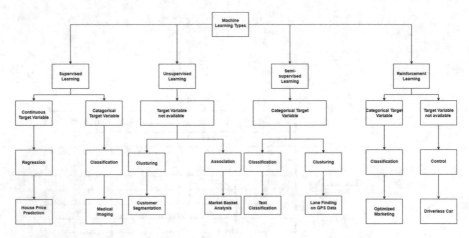

Fig. 2.3 Types of ML algorithms

genuine or fraudulent, whether a sports club will win or lose, whether an earthquake will occur next year or not, weather forecasting and so on. The class is a categorical characteristic separated into levels [27]. A categorical feature known as the class is the target feature to be predicted. There can be two or more levels in a class, and they don't have to be in any specific order. The broad categories of ML algorithms are shown in Fig. 2.3, showing some popular learning algorithms used for various machine learning tasks learning Examples of supervised model construction and prediction is shown in Fig. 2.4 (a) and Fig. 2.4 (b).

2.4.2   Learning Without Supervision

When using unsupervised learning, you have input data (X) and no output variables. In contrast to predictive models that forecast a target of interest, no single feature in a descriptive model is more essential than the others. Unsupervised learning describes the process of training a descriptive model because there is no aim to learn. Although there is no supervision, it can be more challenging to develop applications for descriptive models. Algorithms are left to their own devices to find and present the data's intriguing structure. Clustering and association problems are two types of unsupervised learning challenges [28, 29]. Pattern discovery, for example, is a descriptive modeling process that identifies regular correlations within data. Market basket research using transactional purchase data frequently employs pattern discovery. The purpose here is to identify things that are frequently purchased together, so that the knowledge gained may be used to improve marketing strategies. Unsupervised learning is also useful for email spam filtering, dimensionality reduction, and anomaly identification, among other things [30].

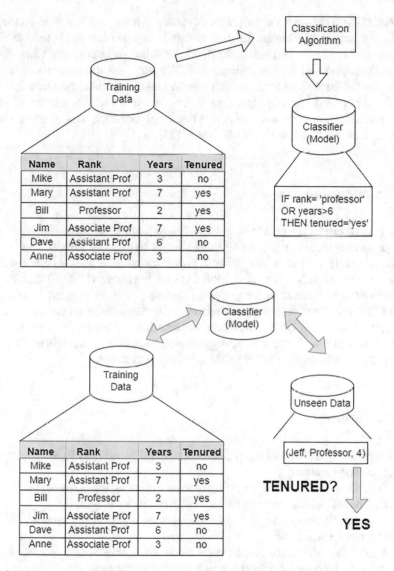

Fig. 2.4 (a) An example of a supervised model construction and prediction. (b) An example of a supervised model construction and prediction

2.4.3 Learning That Is Semi-supervised

Semi-supervised learning combines supervised and unsupervised learning techniques. Only a portion of the data is labeled for the goal or output variable in this type of learning procedure (Y). An excellent example is a photo archive with only a few photographs tagged (e.g. dog, cat, person) and the remainder of the images unlabelled. Labelling data necessitates domain expertise and is a time-consuming and costly operation, whereas unlabelled data is inexpensive and simple to obtain and retain [31, 32]. To find and understand the structure in the input variables, we can use unsupervised learning approaches. We may also use supervised learning approaches to create best guess predictions for unlabelled data, feed that data back into the supervised learning algorithm as training data and then use the model to make predictions on new data that hasn't been seen before. Semi-supervised learning is critical in many real-world ML issues [33].

2.5   Conclusion

A target class can be predicted using a set of input features and effective supervision in a predictive model. A predictive model could be used to forecast past events. Predictive models could adjust traffic signals in real-time during rush hour. The process of constructing a prediction model is known as supervised learning, or more particularly classification and regression. Because there is no goal to learn, the phrase "unsupervised learning" is used to characterise the process of training a descriptive model. However, developing applications for descriptive models can be more difficult because there is no supervision. Algorithms are left to their own devices to identify and portray the data's intriguing structure.

2.6   Exercise

Q1. Describe the differences among "artificial intelligence (AI) vs. data mining vs. ML vs. deep learning".
Q2. Write the various steps of applying machine learning techniques on data.
Q3. "Before ML human learning came" – justify this statement.
Q4. How can we choose a good machine learning algorithm? Explain your view with proper justification.
Q5. Explain the differences among supervised, unsupervised and semi-supervised learning techniques along with some real-life case studies.
Q6. How would you describe ML to a non-technical person?
Q7. What are some examples of AI in use?

References

1. Aristidis Likasa, Nikos Vlassis, Jakob J. Verbeek,"The global k-means clustering algorithm", the journal of the pattern recognition society, Pattern Recognition 36 (2003) 451–461, 2002.
2. Carlos Ordonez, "Clustering Binary Data Streams with K-means", San Diego, CA, USA. Copyright 2003, ACM 1-58113-763-x, DMKD'03, June 13, 2003.

3. K. Wang et al., "A Trusted Consensus Scheme for Collaborative Learning in the Edge AI Computing Domain," in IEEE Network, vol. 35, no. 1, pp. 204–210, January/February 2021, doi:https://doi.org/10.1109/MNET.011.2000249.
4. Guha, D. Samanta, A. Banerjee and D. Agarwal, "A Deep Learning Model for Information Loss Prevention From Multi-Page Digital Documents," in IEEE Access, vol. 9, pp. 80451–80465, 2021, doi:https://doi.org/10.1109/ACCESS.2021.3084841.
5. Rohan Kumar, Rajat Kumar, Pinki Kumar, Vishal Kumar, Sanjay Chakraborty, Prediction of Protein-Protein interaction as Carcinogenic using Deep Learning Techniques, 2nd International Conference on Intelligent Computing, Information and Control Systems (ICICCS), Springer, pp. 461–475, 2021.
6. Guha, A., Samanta, D. Hybrid Approach to Document Anomaly Detection: An Application to Facilitate RPA in Title Insurance. Int. J. Autom. Comput. 18, 55–72 (2021). doi:https://doi.org/10.1007/s11633-020-1247-y
7. Lopamudra Dey, Sanjay Chakraborty, Anirban Mukhopadhyay. Machine Learning Techniques for Sequence-based Prediction of Viral-Host Interactions between SARS-CoV-2 and Human Proteins. Biomedical Journal, Elsevier, 2020.
8. Khamparia, A, Singh, PK, Rani, P, Samanta, D, Khanna, A, Bhushan, B. An internet of health things-driven deep learning framework for detection and classification of skin cancer using transfer learning. Trans Emerging Tel Tech. 2020;e3963. doi:https://doi.org/10.1002/ett.3963
9. Jiawei Han and Micheline Kamber, Data Mining concepts and techniques, Morgan Kaufmann (publisher) from chapter-7 'cluster analysis', ISBN:978-1-55860-901-3, 2006.
10. Dunham, M.H., Data Mining: Introductory And Advanced Topics, New Jersey: Prentice Hall, ISBN-13: 9780130888921. 2003.
11. H. Witten, Data mining: practical machine learning tools and techniques with Java implementations San-Francisco, California: Morgan Kaufmann, ISBN: 978-0-12-374856-0 2000.
12. Kantardzic, M. Data Mining: concepts, models, method, and algorithms, New Jersey: IEEE press, ISBN: 978-0-471-22852-3, 2003.
13. Michael K. Ng, Mark Junjie Li, Joshua Zhexue Huang, and Zengyou He, "On the Impact of Dissimilarity Measure in k-Modes Clustering Algorithm", IEEE transaction on pattern analysis and machine intelligence, vol. 29, No. 3, March 2007.
14. Nareshkumar Nagwani and Ashok Bhansali, "An Object Oriented Email Clustering Model Using Weighted Similarities between Emails Attributes", International Journal of Research and Reviews in Computer science (IJRRCS), Vol. 1, No. 2, June 2010.
15. Oyelade, O. J, Oladipupo, O. O, Obagbuwa, I. C, "Application of k-means Clustering algorithm for prediction of Students' Academic Performance", (IJCSIS) International Journal of Computer Science and Information security, Vol. 7, No. 1, 2010.
16. S. Jiang, X. Song, "A clustering based method for unsupervised intrusion detections" . Pattern Recognition Letters, PP. 802–810, 2006.
17. Steven Young, Itemer Arel, Thomas P. Karnowski, Derek Rose, University of Tennesee, "A Fast and Stable incremental clustering Algorithm", TN 37996, 7th International 2010.
18. Taoying Li and Yan Chen, "Fuzzy K-means Incremental Clustering Based on K-Center and Vector Quantization", Journal of computers, vol. 5, No. 11, November 2010.
19. Tapas Kanungo, David M. Mount, "An Efficient k-Means Clustering Algorithm: Analysis and implementation IEEE transaction vol. 24 No. 7, July 2002.
20. Zuriana Abu Bakar, Mustafa Mat Deris and Arifah Che Alhadi, "Performance analysis of partitional and incremental clustering", SNATI, ISBN-979-756-061—6, 2005.
21. Xiaoke Su, Yang Lan, Renxia Wan, and Yuming, "A Fast Incremental Clustering Algorithm", international Symposium on Information Processing (ISIP'09), Huangshan, P.R. China, August-21-23, pp: 175–178, 2009.
22. Kehar Singh, Dimple Malik and Naveen Sharma, "Evolving limitations in K-means algorithm in data Mining and their removal", IJCEM International Journal of Computational Engineering & Management, Vol. 12, April 2011.
23. Anil Kumar Tiwari, Lokesh Kumar Sharma, G. Rama Krishna, "Entropy Weighting Genetic k-Means Algorithm for Subspace Clustering", International Journal of Computer Applications (0975–8887), Volume 7–No. 7, October 2010.

24. K. Mumtaz, Dr. K. Duraiswamy, "An Analysis on Density Based Clustering of Multi Dimensional Spatial Data", Indian Journal of Computer Science and Engineering, Vol. 1 No 1, pp-8–12, ISSN: 0976-5166.
25. A.M. Sowjanya, M. Shashi, "Cluster Feature-Based Incremental Clustering Approach (CFICA) For Numerical Data", IJCSNS International Journal of Computer Science and Network Security, VOL. 10 No. 9, September 2010.
26. Martin Ester, Hans-Peter Kriegel, Jorg Sander, Michael Wimmer, Xiaowei Xu, "Incremental clustering for mining in a data ware housing", 24th VLDB Conference New York, USA, 1998.
27. Sauravjyoti Sarmah, Dhruba K. Bhattacharyya, "An Effective Technique for Clustering Incremental Gene Expression data", IJCSI International Journal of Computer Science Issues, Vol. 7, Issue 3, No 3, May 2010.
28. Debashis Das Chakladar and Sanjay Chakraborty, Multi-target way of cursor movement in brain computer interface using unsupervised learning, Biologically Inspired Cognitive Architectures (Cognitive Systems Research), Elsevier, 2018.
29. Althar, R.R., Samanta, D. The realist approach for evaluation of computational intelligence in software engineering. Innovations Syst Softw Eng 17, 17–27 (2021). doi:https://doi.org/10.1007/s11334-020-00383-2.
30. B. Naik, M. S. Obaidat, J. Nayak, D. Pelusi, P. Vijayakumar and S. H. Islam, "Intelligent Secure Ecosystem Based on Metaheuristic and Functional Link Neural Network for Edge of Things," in IEEE Transactions on Industrial Informatics, vol. 16, no. 3, pp. 1947–1956, March 2020, doi:https://doi.org/10.1109/TII.2019.2920831.
31. Debashis Das Chakladar and Sanjay Chakraborty, EEG Based Emotion Classification using Correlation Based Subset Selection, Biologically Inspired Cognitive Architectures (Cognitive Systems Research), Elsevier, 2018.
32. D. Samanta et al., "Cipher Block Chaining Support Vector Machine for Secured Decentralized Cloud Enabled Intelligent IoT Architecture," in IEEE Access, vol. 9, pp. 98013–98025, 2021, doi:https://doi.org/10.1109/ACCESS.2021.3095297.
33. CHEN Ning, CHEN An, ZHOU Long-xiang, "An Incremental Grid Density-Based Clustering Algorithm", Journal of Software, Vol. 13, No. 1, 2002.

Chapter 3
Supervised Learning-Based Data Classification and Incremental Clustering

3.1 Introduction

Supervised learning is an ML technique where a learning function is used to predict an output or independent variable (Y) from input variable (X), such that $Y = f(X)$, with the help of supervision of the training labeled data. The algorithm creates predictions on the training data iteratively, and the teacher or supervisor corrects it. The learning is completed when the algorithm achieves a desirable level of performance.

3.1.1 Case Study

Can a diner experience the unseen food for the first time? Someone says "yes" or someone says "no". But, we cannot straightforwardly give this answer. There may be some probability involved while saying "yes" or "no". However, it is fascinating that a diner can recognize an unseen food using his sense of taste, flavour, and smell. There may be a quick data gathering phase at first: what are the most noticeable spices, scents and textures? Is the dish savoury or sweet in flavour? The diner can then use this information to compare the bite to other foods he or she has consumed in the past. Briny flavours may conjure up visions of fish, while earthy flavours may conjure up memories of mushroom-based foods. We think of the discovery process in terms of a slightly modified adage: if it smells like a duck and tastes like a chicken, you're probably eating chicken. This is an example of supervised learning, which is a concept that may be applied to machine learning. Here, collected data and memorizable experiences work like a supervisor who helps make the final decision about the type or name of the food [1, 2]. The supervised learning is broadly classified into two types, classification, and regression. Table 3.1 shows the detailed categorisation of supervised learning techniques.

3.2   Classification

A model or a classifier is built to predict categorical labels in classification, which is a data analysis task. Suppose a bank loan officer is interested in approving a loan request by examining the profile and loan application of the customer and deciding whether it is "safe" or "risky" for the bank. In this example, "safe" or "risky" for the bank loans are predicted categorical labels. Similarly, a medical researcher analyses a patient's tumor and can decide whether the tumor is "malignant" or "benign". In this example, "malignant" or "benign" are predicting categorical labels. These groups can be represented by discrete values, where the order of the values is irrelevant. The data classification task is divided into two steps:

1. There is a predefined set of data classes based on which the training phase of the learning model is built. A training set is made of database tuples (n-dimensional attribute vector) and their associated class labels [3]. The class label attribute is unordered and discrete-valued. It is categorical because each value represents a different category or class [4].
2. The predictive accuracy of the classifier model is measured. Instead of using a training set, we prefer to use a test dataset made up of test tuples and their associated class labels to measure the accuracy of the classifier. These tuples were chosen at random from the overall data set. They are not used to build the classifier since they are not dependent on the training tuples. The percentage of test set tuples appropriately classified by a classifier on a given test set is known as its accuracy. Some widely used classification models are represented in Table 3.1. This section aims to describe all these popular classifiers one by one along with some examples and sample code snippets.

3.2.1   *K-nearest Neighbour Classification*

A K-nearest neighbour (KNN) classification can be used to classify an unknown example with the most common class among K-closest examples. The prime statement of KNN classifier is "tell me who your neighbors are, and it will tell you who you are". KNN method is simple, but it has many powerful applications in the fields

Table 3.1 Categorization of supervised learning techniques

Category	Popular methods
Linear models	Single and multilayer perceptron, linear regression, SVM, SVR, logistic regression
Parametric models	Naïve Bayes, hidden Markov chain models, Gaussian discriminant analysis
Non-parametric models	K-nearest neighbour, Kernel regression, Kernel density estimation
Non-metric models	Decision tree (CART)
Hybrid models	Random forest, bagging, boosting

Fig. 3.1 An example of KNN neighbour

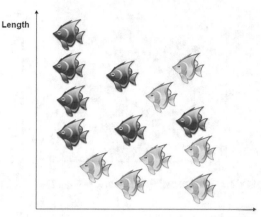

of computer vision, pattern recognition, optical character recognition, facial recognition, genetic pattern recognition, etc. It is also referred to as lazy learner as it waits until the last minute before constructing any model to classify a given test tuple [5, 6]. Unlike eager learners (Bayesian classification, decision trees, etc.), KNN has some limitations, such as the slow classification phase, which requires a large amount of memory, more prone to noisy data or outliers.

In the above Fig. 3.1, there is a troop of sea bass and salmon fishes where you assume $K = 3$ as initial neighbours. Therefore, as shown in the figure, two sea bass and one salmon fish are identified as nearest neighbors. So, your KNN classifier classifies your new input as sea bass. The initial assumption of K will vary. In theory, if infinite numbers of samples are available, the larger is K, the better is classification. The rule of thumb defines $K = \sqrt{n}$, where n is the number of examples, and cross-validation can be used to choose K in some cases. In practice, $K = 1$ is often used for efficiency, but the result can be sensitive to noise. If you choose a large value of K, then it can improve the performance, but too large K destroys the locality. The difference in selecting the different K values has been shown Fig. 3.2.

However, in terms of the dataset, which will be divided into two parts, training and testing, KNN first identifies K-nearest records in the training data using distance-based similarity metrics, and then the unlabelled test records are assigned the class of the majority of the K-nearest neighbours [7].

Illustration-1 (For Numeric Data)

A small student result data table with three attributes (Mathematics, Computer Science, Result), and a set of labelled training samples are given in Tables 3.2 and 3.3 that show an unlabelled testing data sample.

There are two classes of results (pass and fail), and our objective is to determine the actual class (pass or fail) of the given unlabelled test data sample in Table 3.3.

Sol We can use the KNN approach to determine which class better fits the solution. It is also called a binary classification problem as its target class contains only one of the two values. The KNN algorithm treats the attributes or features as coordinates in a multi-dimensional feature space [8]. Our given dataset includes only two

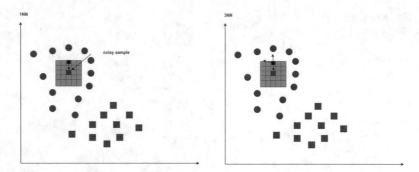

Fig. 3.2 KNN classification with $K = 1$ and $K = 3$ initial neighbours

Table 3.2 A labelled training dataset

SI no.	Mathematics	Computer science	Result (pass/fail)
1	8	7	Pass
2	5	4	Fail
3	9	8	Pass
4	2	4	Fail
5	7	10	Pass

features, and the feature space is two-dimensional. To calculate the nearest distance of two instances or tuples, we can use various distance measures (explained in Chap. 4). However, the most popular distance measure function is Euclidean distance. The below expression can represent Euclidean distance:

$$\text{dist}(p,q) = \sqrt{\left(p_1 - q_1\right)^2 + \left(p_2 + q_2\right)^2 + \ldots + \left(p_n - q_n\right)^2} \tag{3.1}$$

where p and q are the examples to be compared, each having n features. In this problem, we will assume the initial value of K is 3 (an elbow method or $K = \sqrt{n}$ can be considered). So, it is a 3-NN algorithm. There are two steps that we need to follow for the 3-NN classification process:

Step-1: *Find all the K-nearest neighbours of the given testing data sample with the help of distance measure function.*
Step-2: *Apply majority of voting to choose the maximum number of occurrences of the similar class labels.*

SI no.	Euclidean measure	Measured distance	Result
1	$\sqrt{(8-6)^2 + (7-5)^2}$	2.83	*Pass*
2	$\sqrt{(6-5)^2 + (5-4)^2}$	1.4	*Fail*
3	$\sqrt{(6-9)^2 + (5-8)^2}$	4.2	Pass
4	$\sqrt{(2-6)^2 + (4-5)^2}$	4.1	*Fail*
5	$\sqrt{(7-6)^2 + (10-5)^2}$	5.1	Pass

Table 3.3 An unlabelled testing dataset

Mathematics	Computer Science	Result (pass/fail)
6	5	???

Table 3.4 Final testing data with a new class label

Mathematics	Computer Science	Result (pass/fail)
6	5	*Fail*

From the above table, it can be shown that the SI. No. 2, 1, and 4 are the minimum distance-based neighbors where class labels of the two neighbours are fail and one neighbour is pass. So, if we apply the majority of voting technique, we will decide the "fail" as the final class label of the unlabelled sample test data as its frequency of occurrences is higher than the "pass" label [9]. Table 3.4 shows final testing data with a new class label.

Illustration-2 (For Categorical Data)

If our input dataset contains categorical data, we must use the "Hamming distance function" for the distance measure. Hamming distance rule tells that

$$\text{dis}(0,0) = \text{dis}(1,1) = 0$$
and
$$\text{dis}(0,1) = \text{dis}(1,0) = 1$$

Suppose a set of labelled training samples given in Tables 3.5 and 3.6 shows an unlabelled testing data sample. In the training dataset of Table 3.5, there are total four attributes, chills, runny nose, headache and fever and one class label attribute, i.e. flu. It is also a binary classification problem as the target class has two kinds of values such as, no and yes.

However, we will consider the value of K- = 3 and then apply the steps of K-NN algorithm:

Step-1: *Find all the K-nearest neighbours of the given testing data sample with the help of Hamming distance measure function.*

Step-2: *Apply majority of voting to choose the maximum number of occurrences of the similar class labels.*

The calculated Hamming distance for SI. no. 2 in Table 3.7 is evaluated as

$$\text{Dis}(\text{training,testing}) = \left(\begin{array}{l} \textbf{Chills-} > (\text{Yes,Yes}), \textbf{Runny Nose-} > (\textbf{No,Yes}), \\ \textbf{Headache-} > (\textbf{Mild,No}), \textbf{Fever-} > (\textbf{Yes,No}) \end{array} \right) = 3$$

In a similar manner, the distance of the testing sample from all the other training samples has been calculated and given in Table 3.7. However, from Table 3.7, we can finally choose the three minimum distance neighbours as the initial value of

Table 3.5 A labelled training sample dataset

SI no.	Chills	Runny nose	Headache	Fever	Flu
1	Yes	No	Mild	Yes	No
2	Yes	Yes	No	No	Yes
3	Yes	No	Strong	Yes	Yes
4	No	Yes	Mild	Yes	Yes
5	No	No	No	No	No
6	No	Yes	Strong	Yes	Yes
7	No	Yes	Strong	No	No
8	Yes	Yes	Mild	Yes	Yes

Table 3.6 An unlabelled testing dataset

SI no.	Chills	Runny nose	Headache	Fever	Flu
1	Yes	No	Mild	Yes	???

Table 3.7 A training dataset with calculated Hamming distance

SI no.	Chills	Runny nose	Headache	Fever	Flu	Calculated Hamming distance
1	*Yes*	*No*	*Mild*	*Yes*	*No*	*0*
2	Yes	Yes	No	No	Yes	3
3	*Yes*	*No*	*Strong*	*Yes*	*Yes*	*1*
4	No	Yes	Mild	Yes	Yes	2
5	No	No	No	No	No	3
6	No	Yes	Strong	Yes	Yes	3
7	No	Yes	Strong	No	No	4
8	*Yes*	*Yes*	*Mild*	*Yes*	*Yes*	*1*

K = 3. So, SI. no. 1, 3 and 8 are the three minimum distance neighbours, and then we are able to apply the majority of voting on their output class labels (Flu = No, Yes, Yes). Therefore, we will decide the "yes" as the final class label of the unlabelled sample test data as its frequency of occurrences is higher than the "no" label (Table 3.8).

As discussed with respect to Fig. 3.3 above, the accuracy of the KNN classifier will always vary with respect to the initial value of the K. The Fig. 3.3 is representing the relationship of the accuracy (% according to y-axis) results of KNN classifier for the above illustration with respect to the initial values of K (according to x-axis). We can see that when the value of the K is 2 the corresponding accuracy of KNN classifier is 60.34, if K is 5 the accuracy is 78.9, if K is 15 then the accuracy is 91.56 and if K is 20 the accuracy is 62.5. Therefore, it concludes that increasing the number of nearest neighbours (K) values does not always increase the accuracy of the model.

K-nearest neighbour classifier can be very slow when classifying test tuples. If your training dataset has |N| tuples and assume K = 1, then O(|N|) comparisons are needed in order to classify a given test tuple. Parallel implementation or search

Table 3.8 An unlabelled testing dataset

SI no.	Chills	Runny nose	Headache	Fever	Flu
1	Yes	No	Mild	Yes	*Yes*

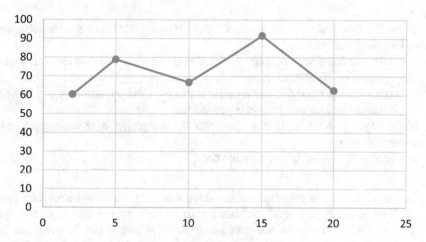

Fig. 3.3 Accuracy(%) of KNN classifier vs. initial values of K

tree-based stored tuple implementation can reduce the overall running time to a constant or O(log|N|), respectively. We can also improve the speed up capability of this classifier by using partial distance calculations based on subset of the n attributes and a predefined threshold [10, 11].

3.2.2 Probabilistic Learning with Naïve Bayes Classification

Naive Bayes performs data classification depends on the Bayes conditional probability theorem. Thomas Bayes introduced this concept. It is a statistical classifier that predicts the probability of a tuple that belongs to a particular class. Probability of an event A can be defined as

$$P(A) = \frac{\text{Total number of events}}{\text{Total number of possible outcomes}}$$

Example 1. If a dice has been thrown (A) and the probability of getting the number 6 is, $P(A) = \dfrac{1}{6}$

Example 2. If a dice has been thrown (A) and the probability of getting the number 2 and number 4 is, $P(A) = \dfrac{1}{3}$.

Naive class conditional independence is a Bayesian concept that states that the influence of one attribute value on a particular class that is independent of the values of the other attributes. As a result, computations are simplified. As an example, a banana is defined as a yellow fruit with a long shape and a 2-inch diameter [12]. These characteristics all contribute to the likelihood that the fruit is a banana. That's why this classifier's name includes the word "naive".

Pros
- Naive Bayes classifier provides high accuracy and speed-up due to the attributes of independence property.
- Naïve Bayes classifier performs well for high-dimensional datasets and multi-class prediction problems.
- This algorithm is quite popular because it can even outperform highly advanced classification techniques. It is an eager learner.
- Naïve Bayes provides a good performance for categorical data in comparison to numerical data.
- It works well with noisy and missing data.

Cons
- In your test data, if a new data comes which was not present in the training data-set, then Naïve Bayes algorithm adds a *zero probability* and fails to do any prediction. This is known as zero-frequency problem, and it can be overcome by smoothing technique (discussed later) [13].
- In real life, it is very difficult to find independent set of features.
- It is not suitable for those datasets with large number of numeric features.

Bayes theorem can be derived from the following expression of conditional probability:

$$P(A \mid B) = \frac{P(A \cap B)}{P(B)}$$

$$\Rightarrow P(A \cap B) = P(B)P(A \mid B)$$

$$\Rightarrow P(B \mid A) = \frac{P(A \cap B)}{P(A)}$$

$$\Rightarrow P(B \mid A) = \frac{P(B)P(A \mid B)}{P(A)}$$

The Bayes theorem is most commonly employed to explain the link between two dependent occurrences. If we think about it, our objective is to estimate the likelihood that an incoming email is spam. $P(B \mid A)$ can be read as the probability of event B given that event A occurred in the last equation. This is known as conditional probability, in which the outcome of event B is determined by the outcome of event A. $P(B)$ is the prior probability, which means that the best guess would be the chance that any previous message was spam, without taking into account any other information. However, you've been taught to look for spam emails that contain an additional phrase, such as Viagra, in the subject line. The likelihood is the probability that the Viagra phrase was used in prior spam messages, and the marginal likelihood is the probability that Viagra appeared in any message at all. We can now compute the posterior probability, which measures how probable the message is to be spam, using the technique above [14, 15]. A threshold value needs to be considered on the posterior probability to decide about the message which more likely to be spam.

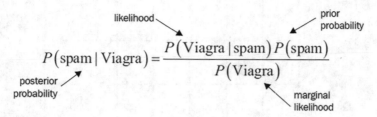

$$P(\text{spam} \mid \text{Viagra}) = \frac{P(\text{Viagra} \mid \text{spam}) P(\text{spam})}{P(\text{Viagra})}$$

There are three types of Naïve Bayes algorithms.

A. *Bernoulli Naïve Bayes*: When the predictors are binary variables (0/1 or yes/no or true/false).
B. *Multinomial Naïve Bayes*: This kind of algorithm is used for the document classification problems with the help of the present word's frequency as features. Suppose multinomial NB algorithm is used to determine whether a document belongs to an "administrative" category or "academic" category in an institute.
C. *Gaussian Naïve Bayes*: When the predictors are continuous in nature, follow the Gaussian distribution pattern.

The basic idea behind a Naïve Bayes classification method is to "determine the probability of a previously unseen instance belonging to each class, and then choose the most likely class". The decision boundary of the Naïve Bayesian Classifier is piecewise quadratic. Before we look at the arithmetic, we can start with a visual perception. The below Fig. 3.4 is the visual example of grasshoppers and katydids classes.

We can analyse the above visual data by build a histogram on antenna length in Fig. 3.5.

Therefore, we can summarise the histogram (Figs. 3.6, 3.7, and 3.8) with two normal distributions. Now, some insect's data have been given whose antenna lengths are 3, 7 and 5 units long, respectively. How can we classify it?

From the above distributions, we can calculate the corresponding probabilities based on the antenna length of the given test sample of an insect.

$P(\text{grasshopper} \mid 3) = 10 / (10+2) = 0.833$
$P(\text{katydid} \mid 3) = 2 / (10+2) = 0.166$
So, it will be a grasshopper with highest probability.

$P(\text{grasshopper} \mid 7) = 3 / (3+9) = 0.250$
$P(\text{katydid} \mid 7) = 9 / (3+9) = 0.750$
So, it will be a katydid with highest probability.

$P(\text{grasshopper} \mid 5) = 6 / (6+6) = 0.500$
$P(\text{katydid} \mid 5) = 6 / (6+6) = 0.500$

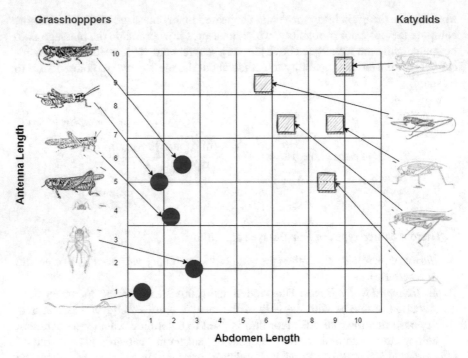

Fig. 3.4 A dotted plot between antenna length and abdomen length of grasshoppers and katydids

Fig. 3.5 A histogram plot between antenna length and abdomen length of grasshoppers and katydids

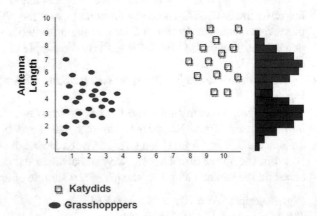

Fig. 3.6 Histogram

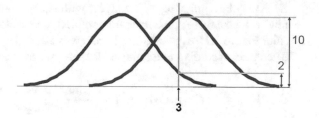

Fig. 3.7 Histogram

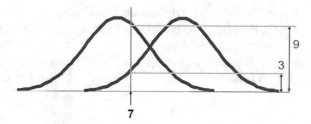

Fig. 3.8 Histogram

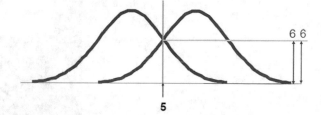

Fig. 3.9 "Drew" can be a male or a female

So, it may be a grasshopper or katydid with equal probabilities. Figure 3.9 shows "Drew" can be a male or a female. Figure 3.10 represents is Officer Drew a male or female? Figure 3.11 expresses Officer Drew is a female.

Therefore, according to the Bayes theorem, we evaluate

$$P(\text{male} \mid \text{drew}) = \frac{P(\text{drew} \mid \text{male}) P(\text{male})}{P(\text{drew})}$$

where P(male|drew) is the probability of being called "Drew" given that he or she is a male, P(male) is the probability of being a male and P(drew) is the probability of

Fig. 3.10 Is Officer Drew a male or female?

Fig. 3.11 Officer Drew is a female

being a name called "Drew". However, we have a small training dataset for the classification and prediction tasks. Our main objective is to identify who is this Officer Drew below.

We can use Bayes rule on the training dataset Table 3.9 to find the probability of drew belongs to a male or female class:

$$P(\text{male|drew}) = \frac{P(\text{drew|male})P(\text{male})}{P(\text{drew})}$$

$$P(\text{male|drew}) = \frac{1/3*3/8}{3/8}$$

$$\Rightarrow \mathbf{P(\text{male|drew}) = 0.125}$$

$$P(\text{female|drew}) = \frac{P(\text{drew|female})P(\text{female})}{P(\text{drew})}$$

$$P(\text{female|drew}) = \frac{2/5*5/8}{3/8}$$

$$\Rightarrow \mathbf{P(\text{female|drew}) = 0.250}$$

The probability of P(female|drew) is greater than P(male|drew). Hence, we can conclude that Officer Drew is more likely to be a female.

So far, we have considered only one attribute or feature to do Bayes classification for the sake of simplicity. However, there can be many features in real-life datasets. We are considering this below dataset Table 3.10 as an example.

Illustration-2

To simplify the task, naïve Bayesian classifiers assume attributes have independent distributions and thereby estimate as

$$P(d|C_j) = P(d_1|C_j) * P(d_2|C_j) * P(d_3|C_j) * \ldots\ldots\ldots * P(d_n|C_j)$$

$P(d|Cj)$ is the probability of class Cj creating instance d, $P(d1|Cj)$ is the probability of class Cj generating the observed value for feature 1, $P(d2|Cj)$, and so on.

$$P(\text{X|Mango}) = P(\text{Sweet|Mango}) * P(\text{Shape|Mango}) = 0.5$$
$$P(\text{X|Banana}) = P(\text{Sweet|Banana}) * P(\text{Shape|Banana}) = 0.2$$
$$P(\text{Sweet|Mango}) = P(\text{Mango|Sweet}).P(\text{Sweet})/P(\text{Mango})$$
$$P(\text{Shape|Mango}) = P(\text{Mango|Shape})P(\text{Shape})/P(\text{Mango})$$

The probability of $P(\text{X|Mango})$ is greater than $P(\text{X|Banana})$. Hence, we can conclude that fruit X is more likely to be a mango rather than banana. Naïve Bayes algorithm has one major limitation. Suppose a new feature enters with the test data, then it is difficult to calculate and use the probability of that new feature ($P(\text{new}) = 0/5$). If that probability is multiplied with other probabilities of the independent features, then the whole term becomes zero. Suppose we have this below dataset:

Table 3.10 A training dataset with many features

Class	Sweet	Shape	Total
Mango	300	540	840
Banana	200	700	900
Total	500	1240	

Fever	Flu	Probability
Hot-2	5	2/5
Mild-3	5	3/5
Severe-0 (new feature)	5	0/5

To overcome this serious limitation in Naïve Bayes algorithm, we have to add "1" to the numerator part and add number of levels "3" (hot, mild and severe) in the denominator part. This process is called *add-1 smoothing* of NB algorithm.

Fever	Flu	Probability
Hot-2+1	5+3	3/8
Mild-3+1	5+3	4/8

Table 3.9 Training dataset based on Bayes

Class	Flu	Probability
Male	5	2/5
Female	5	3/5
Kids	5	0/5

Fever	Flu	Probability
Severe-0+1 (new feature)	5+3	1/8

The Naïve Bayes classifier makes the assumption of class conditional independence, which means that the values of the features in a particular tuple's class label are expected to be conditionally independent of one another. If this assumption is correct, the NB becomes more effective than any other classifier.

3.2.3   *Divide and Conquer Classification*

3.2.3.1   **Rule-Based Learning**

This classifier divides data into smaller chunks in order to find patterns that can be used to make future predictions. This machine learning method represents knowledge as logical structures, which is very valuable for business planning and process development. The trained model is expressed as a series of IF-THEN rules in a rule-based classifier. An IF-THEN rule can be expressed in the form as

$$IF\ condition\ \textbf{THEN}\ conclusion$$

Example:

Rule : **IF** $age = old\ AND\ blood - pressure = high\ \textbf{THEN}\ heart - attack = yes$

IF part holds precondition statement and THEN part holds consequent. A pre-condition statement consists of one or more attributes along with some logical connectives (AND, OR, etc.). If the precondition for all the attributes holds true for a given tuple, then we can conclude that the rule consequent is satisfied and covers the tuple. The fraction of tuples covered by a rule (i.e. whose attribute values hold true for the rule's antecedent) is called coverage. To determine the correctness of a rule, we examine the tuples it covers and determine what percentage of them the rule can correctly classify:

$$coverage = \frac{Ncov}{|D|}$$

$$accuracy = \frac{Ncorr}{Ncov}$$

Here, Ncov is the number of tuples covered by the rule, Ncorr is the number of tuples correctly classified by the rule and |D| be the total number of tuples in the dataset D. Suppose we consider the below dataset table.

Age	Blood pressure	Heart attack
Young	Low	No
Young	High	No
Old	Low	No
Old	High	Yes

Our main objective is to predict whether a patient will face heart attack or not. Consider the above rule, which covers one tuple out of four tuples. It can correctly classify that one tuple. Therefore, *coverage*(Rule) = 1/4 = 25% and *accuracy*(Rule) = 1/1 = 100%.

3.2.3.2   Decision Tree-Based Learning

Learners who use decision trees create a model in the form of a flowchart-like tree structure. It's a method of learning based on class-labelled training tuples. Each decision tree has a decision node that represents a decision to be made on an attribute, and the tree's edges represent the decision's options. Each of the decision tree's terminal nodes (Figs. 3.12 and 3.13) represents a class label [16]. The root node is the highest node in the tree. Data categorisation starts at the root node and

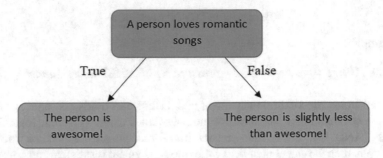

Fig. 3.12 The decision tree's terminal nodes

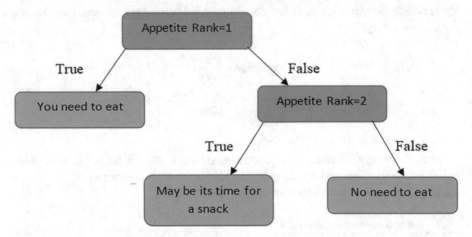

Fig. 3.13 The decision tree's terminal nodes

progresses through a series of tests based on the values of the features or attributes. Credit scoring models in which the reasons that cause an applicant to be denied must be well-specified are a few potential uses of the decision tree classification.

The above decision tree is very simple. It asks a question (yes/no) whether the person loves to listen to romantic songs, and then it classifies the person based on the answer.

This decision tree is based on a ranked data where 1 is super hungry and 2 is moderately hungry. If a person is super hungry, they need to eat; if a person is moderately hungry, they just need a snack, and if they are not hungry at all then they don't need to eat. The classification can be categories or numeric. A little bit complex decision tree example is shown in Fig. 3.6. It represents the concept of getting loan approval where it predicts whether a customer is likely to receive the loan. Some decision tree algorithms create binary trees solely, whereas others create non-binary trees. When an unknown tuple with an unknown class label is received, the tuple's attribute values are compared to the decision tree. As a result, a dedicated path from the root to a specific leaf node is traced in order to forecast the class label

Fig. 3.14 A decision tree
where each internal node
represents a test on
attribute and each leaf
node represents a class

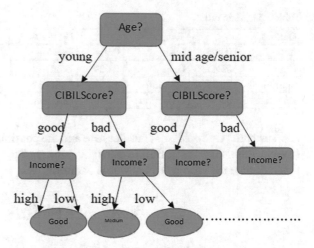

of that unknown tuple. Decision trees make it simple to derive classification rules.
Figure 3.14 shows decision tree where each internal node represents a test on attribute and each leaf node represents a class.

The decision tree classifier is popular because it requires minimal domain knowledge and can handle large amounts of data. The representation of learned knowledge in the form of a tree structure is straightforward and simple to comprehend.
This method's learning and classification are both quick and accurate. Depending
on the nature of the supplied data, it also delivers good accuracy [17, 18]. Many of
the branches in decision trees may reflect noise or outliers in the training data. Tree
trimming aims to locate and delete such branches in order to improve classification
accuracy on data that isn't visible. A big number of numeric features can sometimes
lead to a huge number of decisions and an unduly complicated tree. It uses top-
down recursive partitioning approach as it splits the attribute values into smaller
subset of similar classes. Some famous decision tree algorithms are ID3, C4.5 and
Classification and Regression Tree (CART). All these algorithms follow a greedy
approach. In this chapter, we prefer to discuss about the Classification and Regression
Tree (CART) algorithm.

Given training set D, the algorithm's computational complexity is $O(n|D|\log(|D|))$,
where n is the number of attributes characterising the tuples in D and |D| is the num-
ber of training tuples in D. With |D| tuples, this indicates that the computational cost
of constructing a tree grows at most $n|D|\log(|D|)$. The reader is left with the proof as
an exercise. There have also been proposals for incremental versions of decision
tree induction. Instead than learning a new tree from scratch when given new training data, they reconstruct the decision tree they learned from earlier training
data [19].

The input dataset has been given in Table 3.11. It contains mainly four attributes
(Outlook, Temp, Humidity and Winds) and one class attribute "Play Tennis". It is a
binary classification problem. However, we build the decision tree here with the
help of evaluation "information gain (IG) or entropy" of each given attribute. So, the

Table 3.11 Gain ratio

Outlook		Temp		Humidity		Winds	
Info	0.693	Info	0.911	Info	0.788	Info	0.892
Gain	0.247	Gain	0.029	Gain	0.152	Gain	0.048
Splitinfo	1.577	Split info	1.557	Split info	1.000	Splitinfo	0.985
Gain ratio	0.157	Gain ratio	0.019	Gain ratio	0.152	Gain ratio	0.049

first attribute "Outlook" we consider here and the possible values of the Outlook are {Sunny, Rain, Overcast}. Before calculating the entropy of any attribute, we have to first calculate the entropy of the whole dataset:

$$\text{Entropy} = \sum_{i=1}^{n} P_i \log_2 (P_i)$$

So, in the whole dataset (S), the total number of yes (P) (positive classes) is 9, and the total number of No (N) (negative classes) is 5:

$$\text{InformationGain}_S = -\frac{P}{P+N} \log_2 \left(\frac{P}{P+N}\right) - \frac{N}{P+N} \log_2 \left(\frac{N}{P+N}\right)$$

$$\text{InformationGain}_S = -\frac{9}{14} \log_2 \left(\frac{9}{14}\right) - \frac{5}{14} \log_2 \left(\frac{5}{14}\right)$$

$$\text{Entropy or IG}_S (P,N) = 0.94$$

Therefore, we will calculate the entropy of the first attribute Outlook with respect to Sunny, Overcast and Rain:

$$\text{IG}_S (\text{Sunny}) = -\frac{2}{5} \log_2 \frac{2}{5} - \frac{3}{5} \log_2 \frac{3}{5}$$
$$= 0.970$$

Outlook	P_i	N_i	$IG_S(P_i,N_i)$
Sunny	2	3	0.970
Overcast	4	0	0
Rain	3	2	0.970

$$\text{Entropy}(\text{Outlook}) = \sum_{i=1}^{v} \frac{P_i + N_i}{P+N} \left(IG_S (P_i,N_i)\right)$$

$$\text{Entropy}(\text{Outlook}) = \frac{5}{14} \times 0.970 + \frac{4}{14} \times 0 + \frac{5}{14} \times 0.970$$

$$\text{Entropy}(\text{Outlook}) = 0.692$$

So, we can now calculate the gain of Outlook,

Gain(Outlook) = IG_S – Entropy(Outlook) = **0.248(Maximum)**

Therefore, we will calculate the gain of other attributes (Temp, Humidity and Winds) in similar ways as follows:

Gain(Temp) = 0.029
Gain(Humidity) = 0.151
Gain(Winds) = 0.048

From the above explanation, it is clearly seen that the attribute Outlook has the maximum gain, so we will consider Outlook as the root node of the decision tree for this problem. There are three categories of the Outlook attribute (Sunny, Overcast and Rain), so our root node must have three sub-nodes. The decision tree with some known and unknown decision has been shown below. According to the Table 3.11, the Sunny component of attribute Outlook consists of D_1, D_2, D_8, D_9 and D_{11} tuples where the target class labelled as 2 positive and 3 negative, so we are not able to take any decision. Whereas, for Overcast scenario, there are 4 positive and 0 negative tuples exist, so we can easily take decision as positive (yes) for the target class. Rain component also have 3 positive and 2 negative decisions. Figure 3.15 shows decision tree after the attribute selection for the root node at Level-1.

{D1,D2....,D14} (9P,5N)

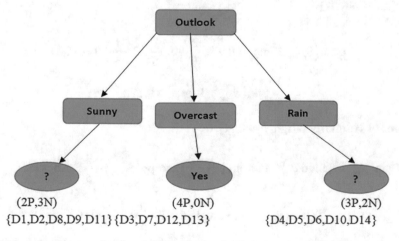

Fig. 3.15 A decision tree after the attribute selection for the root node at Level-1

Now, we have taken all the five example tuples represent Sunny components.

Day	Temp	Humidity	Winds	Play tennis
D1	Hot	High	Weak	No
D2	Hot	High	Strong	No
D8	Mild	High	Weak	No
D9	Cool	Normal	Weak	Yes
D11	Mild	Normal	Strong	Yes

Now, we will calculate the gain of all three attributes (Temp, Humidity, Winds) of the above table and then build a decision tree.

Values of Temp {Hot, Mild, Cool}:

$$\text{Entropy}\left(S_{\text{Sunny}}\right) = -\frac{2}{5}\log_2\frac{2}{5} - \frac{3}{5}\log_2\frac{3}{5}$$
$$= 0.97$$

Temp	P_i	N_i	$IG_S(P_i,N_i)$
Hot	0	2	0.0
Mild	1	1	1.0
Cool	1	0	0.0

$$\text{Gain}\left(S_{\text{sunny}}, \text{Temp}\right) = \text{Entropy}\left(S\right) - \frac{2}{5}\text{Entropy}\left(S_{\text{Hot}}\right) - \frac{2}{5}\text{Entropy}\left(S_{\text{Mild}}\right)$$
$$-\frac{1}{5}\text{Entropy}\left(S_{\text{Cool}}\right)$$

$$\text{Gain}\left(S_{\text{sunny}}, \text{Temp}\right) = 0.97 - \frac{2}{5}0.0 - \frac{2}{5}1.0 - \frac{1}{5}0.0 = 0.570$$

$$\textbf{Gain}\left(S_{\textbf{Sunny}}, \textbf{Temp}\right) = \textbf{0.570}$$

Values of Humidity {High, Normal}:

$$\text{Gain}\left(S_{\text{sunny}}, \text{Humidity}\right) = \text{Entropy}\left(S\right) - \frac{3}{5}\text{Entropy}\left(S_{\text{High}}\right) - \frac{2}{5}\text{Entropy}\left(S_{\text{Normal}}\right)$$

$$\text{Gain}\left(S_{\text{sunny}}, \text{Humidity}\right) = 0.97 - \frac{3}{5}0.0 - \frac{2}{5}0.0$$

$$\textbf{Gain}\left(S_{\textbf{Sunny}}, \textbf{Humidity}\right) = \textbf{0.97}\,(\textbf{Maximum})$$

Values of Winds {Strong, Weak}:

$$\text{Gain}\left(S_{sunny}, \text{Winds}\right) = \text{Entropy}(S) - \frac{2}{5}\text{Entropy}\left(S_{Strong}\right) - \frac{3}{5}\text{Entropy}\left(S_{Weak}\right)$$

$$\text{Gain}\left(S_{Sunny}, \text{Winds}\right) = 0.97 - \frac{2}{5}1.0 - \frac{3}{5}0.918$$

$$\textbf{Gain}\left(\textbf{S}_{\textbf{Sunny}}, \textbf{Winds}\right) = \textbf{0.0192}$$

Due to the maximum gain, we will consider "Humidity" as a node for the next level of left subtree. Figure 3.16 shows decision tree after attribute selection for the root node of left subtree at Level-3

From the above decision tree, we can derive that whenever Humidity is "high", then the three tuples $\{D_1, D_2, D_8\}$ represent "no" as decision and when Humidity is "normal", then the two tuples $\{D_9, D_{11}\}$ represent "yes" as decision. But, still the Rain component has no target decision yet. So, we consider the below table to evaluate it.

Day	Temp	Humidity	Winds	Play tennis
D4	Mild	High	Weak	**Yes**
D5	Cool	Normal	Weak	**Yes**
D6	Cool	Normal	Strong	No
D10	Mild	Normal	Weak	**Yes**
D14	Mild	High	Strong	No

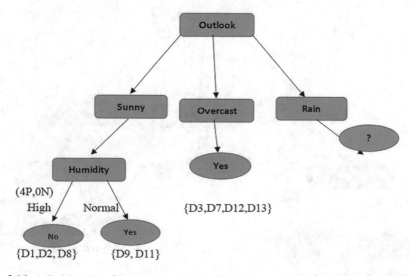

Fig. 3.16 A decision tree after attribute selection for the root node of left subtree at Level-3

Values of Temp {Hot, Mild, Cool}:
Entropy $(S_{Rain}) = 0.97$
Entropy $(S_{Hot}) = 0.0$
Entropy $(S_{Mild}) = 0.9183$
Entropy $(S_{Cool}) = 1.0$
Gain $(S_{Rain,}$ Temp) = 0.0192
Values of Humidity {High, Normal}:
Entropy $(S_{High}) = 1.0$
Entropy $(S_{Normal}) = 0.9183$
Gain $(S_{Rain},$ Humidity) = 0.0192
Values of Winds {Strong, Weak}:
Entropy $(S_{Strong}) = 0.0$
Entropy $(S_{Weak}) = 0.0$
Gain $(S_{Rain},$ Winds) = 0.97 (Maximum)

Due to the maximum gain, we will consider "Winds" as a node for the next level of left subtree. Figure 3.17 shows decision tree after attribute selection for the root node of right subtree at Level-3.

From the above decision tree, we can derive that whenever Winds is "strong", then the two tuples $\{D_6, D_{14}\}$ represent "no" as decision and when Winds is "weak", then the three tuples $\{D_4, D_5, D_{10}\}$ represent "yes" as decision. This tree is our final decision tree that we have drawn by following the ID3 procedure.

Gain Ratio
The notion of gain ratio was introduced in the next modified decision tree classifier C4.5, which is the successor to the ID3 method. Information gain, often known as entropy, is an extension of gain ratio. Ross Quinlan had proposed it. Entropy favours

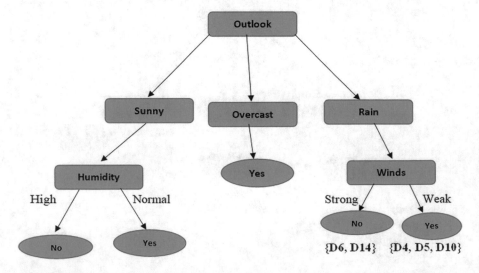

Fig. 3.17 A decision tree after attribute selection for the root node of right subtree at Level-3

tests with a large number of outcomes [20, 21]. That example, selecting properties with a small number of values is not recommended. Because an employee Id is a unique identifier, dividing on this attribute will result in a high number of partitions, providing the most information. When choosing an attribute, the gain ratio considers the number and size of branches. A normalisation of entropy will be used to compute the gain ratio:

$$\text{Splitinfo}_{\text{X}}(\text{D}) = \sum_{j=1}^{\upsilon} \frac{|\text{D}_j|}{|\text{D}_v|} \times \log_2 \left(\frac{|\text{D}_j|}{|\text{D}_v|} \right)$$

Where v is the number of partitions of the training dataset D and A is the testing attribute. The gain ratio is defined as

$$\text{Gain ratio}(\text{A}) = \frac{\text{Gain}(\text{A})}{\text{Split in of}(\text{A})}$$

Gain ratio for the Table 3.11 is the following.

The attribute with the highest gain ration will be considered as the splitting attribute. From the above table, we can see that the attribute "Outlook" still comes out top to be acted like a root node.

Cons
- It only considers the attributes with greater than the average information gain.
- It may over compensate.

Gini Index
Gini index concept is introduced along with the CART algorithm. It is also called as Gini impurity measures that specify the probability of a particular variable being misclassified when it is selected randomly. It can be calculated by subtracting the sum of the squared probabilities of tuples in the dataset D that belongs to each class from the integer value 1. It can be represented for m number of classes as

$$\text{Gini}(\text{D}) = 1 - \sum_{i=1}^{m} P_i^2,$$

$$\text{Gini}(\text{D}) \in 0,1$$

When Gini index is "0", that means all the tuples belong to a single class and "1" indicates tuples are randomly distributed among different classes. While building the decision tree, we would prefer selecting the attribute with the least Gini index as the root node. Unlike entropy, Gini index does not include logarithmic function for calculation.

Pros of Decision Tree Algorithm
1. It is a general classifier, which performs well for most of the problems.
2. It is capable of handling missing data and numeric features.
3. It can be applied on dataset that has relatively few training examples.
4. For building a relatively small tree model, it does not require any mathematical background.

5. Easy to understand and more efficient than other complex models.

Cons of Decision Tree Algorithm
1. It is often biased towards splits on features when it interprets a large number of levels.
2. Overfitting and underfitting are very common problems for this kind of model.
3. A very small change in training data can make a large impact on the decision logic.
4. Sometimes, large decision trees are difficult to handle and interpret.

We can improve the efficiency of the decision tree algorithm by doing tree pruning.

3.3   Decision Tree Pruning

Decision tree is prone to overfitting the training data as it grows indefinitely by dividing into smaller partitions with the help of splitting attributes process. Pruning is mainly responsible to prevent overfitting to noise in the data [21]. There are two kinds of pruning techniques available"

- **Prepruning:** When information is unreliable, it will halt growing a branch.
- **Postpruning:** It takes a fully formed decision tree and prunes out the portions that aren't reliable.

In practice, postpruning is favoured over prepruning since prepruning can stop too early, and it is impossible to identify the appropriate depth of a decision tree without first growing it.

3.4   Postpruning Operations

It entails developing an overly large tree and then employing pruning criteria based on node error rates to lower the tree's size to a more acceptable level.

1. Create a comprehensive decision tree that overfits the training data first.
2. Nodes and branches that have a minor impact on categorisation errors are deleted afterwards.
3. Then prune in one of two ways: (1) subtree replacement or (2) subtree raising.
4. Subtree replacement (bottom-up technique) involves moving or replacing branches farther up the tree. It only considers removing a tree after taking into account all of its subtrees.
5. In subtree raising approach, a node can be deleted or redistribute instances. But this approach is slower than subtree replacement.

Prepruning Operations
When there are no statistically significant correlations between any characteristic and the class at a given node, it can cease constructing the tree. The Chi-squared test is the most commonly used statistical test for this purpose [22].

3.5   Random Forests Classification

The decision tree classification method is inflexible and yields erroneous results. Random forests, an ensemble learning method for classification, were created to improve data accuracy. Random forests are built up of decision trees and are an extension notion of decision trees. It combines the flexibility of decision trees with the simplicity of decision trees, resulting in significant increases in accuracy. Leo Breiman and Adele Cutler championed this strategy, which combines the basic concepts of bagging with random feature selection to bring more variation to decision tree models [23].

Procedure
- Create a bootstrapped dataset in step one. The size of the bootstrapped dataset is the same as the original. We simply took samples from the original dataset at random and saved them in the bootstrapped dataset. We are permitted to select the same sample many times.
- Step 2: Using the bootstrapped dataset, create a decision tree with only a random selection of variables at each step.
- Step-3: Return to Step 1 and repeat: create a fresh bootstrapped dataset, and build a new decision tree at each step using a subset of variables. It produces a wide range of trees. Random forest is more successful than individual decision trees because of its variety.
- Step 4: Add the new unidentified sample data to each of the tree varieties. We'll take the majority of votes and make the final choice or prediction about the target class after processing the data through all of the trees in random forest.

Instead of considering all four variables to figure out how to split the root node, we consider here only two variables randomly. In this case, we randomly select "blood circulation" and "blocked arteries" as candidates for the root node. For the sake of best job, we consider "blood circulation" as the root node of our first decision tree. Now, we have to choose the next split under this root node among the three remaining variables randomly (chest pain, blocked arteries and weight). So, we build the tree as usual but only considering a random subset of variables at each step [24]. Figure 3.18 represents sample decision tree developed from the first bootstrapped dataset.

We can now return to step one and continue the process by creating a fresh bootstrapped dataset and building a new tree by examining a subset of variables at each stage. In practice, using the random forests procedure 100 times produces a vast

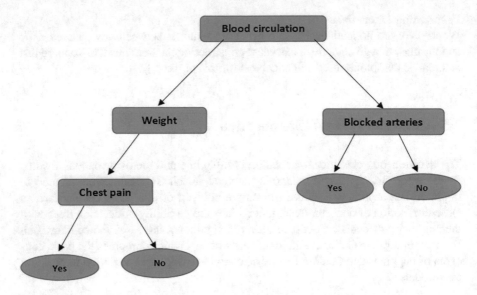

Fig. 3.18 A sample decision tree developed from the first bootstrapped dataset

diversity of trees. Random forests are more successful than individual decision trees because of their variety.

Testing Phase
Now, we receive a new sample as a test data of a new patient.

Chest pain	Blood circulation	Blocked arteries	Weight	Heart disease
Yes	No	No	168	?

Now, we run this test sample data in our first decision tree and it concludes "yes" the patient has the heart disease. We keep track of it.

Heart disease	
Yes	No
1	0

Now, we run the same data in the second decision tree that will also conclude with "yes", and we update our record accordingly.

Heart disease	
Yes	No
2	0

In this way, we follow the same approach and run the sample test data for all the other variants of the decision trees. For sake of simplicity, if we assume that we have total six variants of the decision trees for the above dataset and after running the sample test data, the final decision table will look like below where five decision trees says "yes" and one says "no".

Heart disease	
Yes	No
5	1

In this case, the "yes" received the most votes (majority of voting), and we will conclude that this patient has heart disease. Bootstrapping the data plus using the aggregate to make a decision is called "bagging". As we have duplicate entries in the bootstrapped dataset, as a result, the third instance was not included as an instance of a bootstrapped dataset. Typically, about 1/3 of the original data does not end up in the bootstrapped dataset. The below entry did not end up in the bootstrapped dataset [25].

Chest pain	Blood circulation	Blocked arteries	Weight	Heart disease
Yes	Yes	No	210	No

If the original dataset is larger, then we could have more than one entry above. This is called the "Out-of-Bag" dataset. Since this "Out-of-Bag" dataset was not used to create this tree, we can run this data through the tree and notice if it correctly classify the sample as "no" heart disease. Suppose this tree is correctly classify the "Out-of-Bag" sample as "no", then we run this "Out-of-Bag" sample through all the other trees which were built without it. As a result, some trees are incorrectly labelled the "Out-of-Bag" sample as "yes", and some tree does it correctly. Since the label with the most votes wins, it is the label that we assign this "Out-of-Bag" sample. In this case, the "Out-of-Bag" sample is correctly labelled by the random forest. We can apply this for all the other "Out-of-Bag" samples (if we have). Finally, the proportion of "Out-of-Bag" samples correctly identified by the random forest may be used to determine how accurate our random forest is. The percentage of improperly identified "Out-of-Bag" samples is known as "Out-of-Bag" error [26].

Pros
1. An all-purpose model capable of handling almost all the problems.
2. It is capable of handling noisy or missing data and continuous features.
3. It can be used for the dataset which has a large number of input features.

Cons
1. The model cannot be easily interpretable, and it also requires some efforts to tune the model.

3.6   Support Vector Machine Classification

In the era of 1990s, the support vector machine (SVM) classification approach was introduced and well accepted. SVM is a comparatively new promising classifier that suitable for both linear and non-linear data classification. The SVM concept is an extension of support vector classifier that is also a further extension of maximal margin classifier. However, maximal margin classifiers are not able to classify non-linear class boundaries. SVM is mainly intended for the binary classification, but an upgraded SVM also supports multi-class classification of data.

Maximal margin classifier is fully dependent on the concept of separating hyperplanes that is farthest from the training observations. A hyperplane is flat affine subspace of dimension $(m-1)$ in a m-dimensional space. For example, a hyperplane is a simple straight line in a flat one-dimensional subspace, whereas it is a plane of two $(m = 3-1)$ dimensions in a flat three $(m = 3)$-dimensional subspace. In more than three dimensions, it is working like a hyperplane that is very hard to visualise. In two dimensions, a hyperplane can be mathematically defined as

$$\alpha_0 X_0 + \alpha_1 X_1 = 0$$

where $X = (X_1, X_2)$ holds a point on the hyperplane. For m-dimensional space, the hyperplane can be expressed as

$$\alpha_0 X_0 + \alpha_1 X_1 + \alpha_2 X_2 + \ldots \ldots + \alpha_m X_m = 0$$

Now, suppose X does not satisfy the above equation, and then

$$\alpha_0 X_0 + \alpha_1 X_1 + \alpha_2 X_2 + \ldots \ldots + \alpha_m X_m > 0$$

or

$$\alpha_0 X_0 + \alpha_1 X_1 + \alpha_2 X_2 + \ldots \ldots + \alpha_m X_m < 0$$

The above two equations tells that X lies on two different sides of the hyperplane, respectively. So, it can be easily determine on which side of the hyperplane the point located by inspecting the sign of the above equations. However, the problem is to draw an infinite number of hyperplanes in this case. Figure 3.19 says four separating hyperplanes that separate two classes of observations.

The main problem now is deciding one of the infinitely many different separating hyperplanes to employ to build a classifier. To do so, we must first compute the perpendicular distances between each training observation and a specified separating hyperplane and then choose the lowest of these distances, which is known as margin [27]. The maximal margin hyperplane is the separating hyperplane with the biggest margin. Each class's closest observations should be well spaced from the decision boundary. We can classify the test observations based on the separating hyperplane margin once the maximal margin has been determined. However, while this type of classifier is effective, it is prone to overfitting when the m is large. The maximal margin classifier can be constructed as

Fig. 3.19 Four separating
hyperplanes that separate
two classes of observations

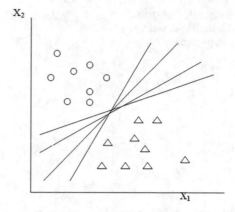

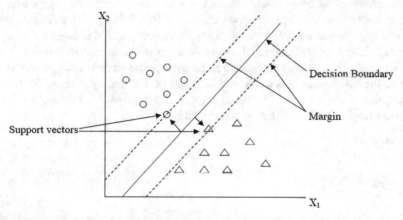

Fig. 3.20 An example of maximal margin hyperplane

Maximise M

$\alpha_0, \alpha_1, \alpha_2,....,\alpha_m$

Subject to $\sum_{j=1}^{m} \alpha_j^2 = 1$

$$y_i \left(\alpha_0 + \alpha_1 X_i 1 + \alpha_2 X_i 2 + \ldots\ldots + \alpha_m X_i m \right) \geq M, \text{for all } i = 1, 2, 3, \ldots\ldots, n$$

M defines the margin of the hyperplane, and we have to choose $\alpha_0, \alpha_1, \alpha_2,....,\alpha_m$ to maximise *M*. Figure 3.20 says an example of maximal margin hyperplane.

Maximal margin classifier fails if there is no separating hyperplane exist, i.e. the optimisation problem has no solution with *M* > 0. This margin classifier is further divided into two groups, (i) hard margin classifier and (ii) soft margin classifier. Figure 3.21 shows two non-separable classes by hyperplane.

$$y_i \left(\alpha_0 + \alpha_1 X_i 1 + \alpha_2 X_i 2 + \ldots\ldots + \alpha_m X_i m \right) \geq M, \text{for all } i = 1, 2, 3, \ldots\ldots, n$$

Fig. 3.21 Two non-separable classes by hyperplane

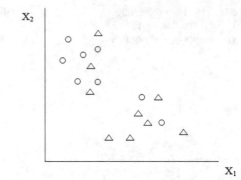

The above equation represents the concept of hard margin classifier where it is impossible to separate the above two classes of Fig. 3.13 by any stringent hyperplane. Tiny margin separation is also not well accepted. However, there is one approach called the soft margin classifier or support vector classifier which is very flexible to allow some observations to be present on the incorrect side of the margin or even the hyperplane. It is called soft as it can be violated by some of the training observations. The wrong side observations of the hyperplane are misclassified by the soft margin classifier. It can be expressed as

Maximise M

$\alpha_0, \alpha_1, \alpha_2,.....,\alpha_m$

Subject to $\sum_{j=1}^{m}\alpha_j^2 = 1$

$$y_i\left(\alpha_0 + \alpha_1 X_i 1 + \alpha_2 X_i 2 + ... + \alpha_m X_i m\right) \geq M\left(1-\varepsilon_i\right), \varepsilon_i \geq 0, \sum_{i=1}^{n}\varepsilon_i \leq C$$

for all $i = 1,2,3,......,n$

where C is a positive tuning parameter, M is the largest width of the margin and ϵ_i is a collection of slack variables that represent errors generated due to the observations to be on the wrong side of the hyperplane. If $\epsilon_i = 0$, the ith test observation belongs to the correct side of the hyperplane, and if $\epsilon_i > 0$ or $\epsilon_i > 1$, then the ith observation is on the wrong side of the hyperplane.

So, we can classify the new test observation (x') as follows:

$$f\left(x'\right) = \alpha_0 + \alpha_1 X_1' + \alpha_2 X_2' + ... + \alpha_m X_m'$$

Therefore, the support vector machine (SVM) classifier was introduced, as it is much less prone to overfitting than other methods. SVM is capable to use for both the classification and numeric predictions. For example, SVM can be used for the classification of microarray gene expression data in bioinformatics or genetic engineering, text categorisation and document classification, natural language processing, detection of earthquakes and combustion engine failure, etc. Traditionally,

SVM is effectively used for binary classification and for multi-class classification SVM adapts some new strategies (one vs. one and all vs. one) [28, 29]. SVM is quite capable to convert a linear classifier into one that produces non-linear decision boundaries. An example of a dataset that needs a non-linear boundary for classification is shown in Fig. 3.14. If we apply support vector classifier, which seeks for linear boundary for this classification, we generate very poor results (Fig. 3.15). In such cases, we can choose support vector machine (SVM). In SVM, we can enlarge the feature space using quadratic, cubic or any higher order polynomial functions of the predictors to support this non-linearity issue:

Maximize M

$$\alpha_0, \alpha_{11,} \alpha_{12}, \ldots\ldots, \alpha_{m1}, \alpha_{m2,} \varepsilon_{1,} \ldots\ldots\ldots, \varepsilon_m$$

$$\text{Subject to,} \quad y_i \left(\alpha_0 + \sum_{j=1}^{m} \alpha_{j1} x_{ij} + \sum_{j=1}^{m} \alpha_{j2} x_{ij}^2 \right) \geq M \left(1 - \varepsilon_i \right),$$

$$\sum_{i=1}^{n} \varepsilon_i \leq C_i, \varepsilon_i \geq 0, \sum_{j=1}^{m} \sum_{k=1}^{2} \alpha_{jk}^2 = 1$$

In enlarged feature space, the above equation works like a linear boundary, but in original space, it will work like a non-linear quadratic polynomial decision boundary. Figure 3.22 shows a dataset that requires non-linear boundary for classification. Figure 3.23 shows results generated after applying support vector classifier.

As a result, SVM is nothing more than a technique for extending the feature space used by the support vector classifier in such a way that calculations are faster.

Fig. 3.22 A dataset that requires non-linear boundary for classification

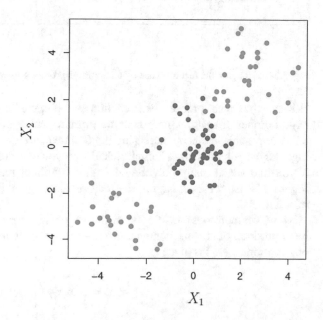

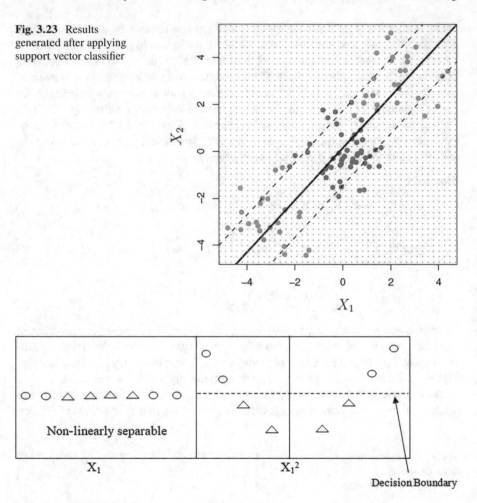

Fig. 3.23 Results generated after applying support vector classifier

Fig. 3.24 Enlarging the feature space by adding higher order polynomial kernels

However, we must proceed with caution while expanding the feature space, as a huge number of features may become unmanageable in practice. Using kernels, SVM may achieve this by increasing the feature space quality [30]. Even with very high degree polynomials, a kernel trick allows you to accomplish the same effect as if you had added many polynomial features without really having to add them. Figure 3.24 expresses enlarging the feature space by adding higher order polynomial kernels.

Now, we represent the solution of the support vector classifier with the help of inner products of the observations, in a feature space, an inner product between two observations (x_i, x_i') is given as

$$x_i.x_i' = \sum_{j=1}^{m} x_{ij} x_{i'j}$$

An SVM with radial kernel can be expressed as

$$Kx_i, x_{i'} = \exp\left(-\gamma \sum_{j=1}^{m} \left(x_{ij} - xi'j \right)^2 \right)$$

where γ is a positive constant. When the value of γ is high, then it considers only the nearby points or objects, and when the γ is low, far-away points are also considered. Figure 3.25 shows an SVM with polynomial kernel of $d = 3$ (left side) and an SVM with radial kernel (right side) applied.

3.7   Multi-class SVM

SVM is typically applicable for the binary classification kind of problems. However, two most popular strategies of extending SVM for the multi-class problems have been accepted.

(i)   One vs. One SVM Classification

If there are k numbers of classes, then there are k*(k−1)/2 pairs of classes for comparisons. For example, we have three classes (Red, Green and Blue), and we have to apply multi-class SVM approach and get the predicted output as

$$SVM\left(Red, Green \right) = Red$$
$$SVM\left(Green, Blue \right) = Green$$
$$SVM\left(Red, Blue \right) = Red$$

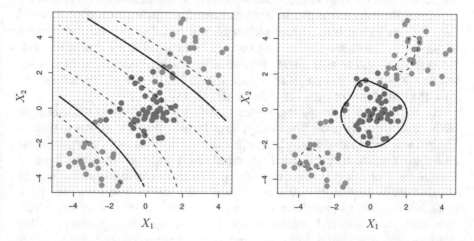

Fig. 3.25 An SVM with polynomial kernel of $d = 3$ (left side) and an SVM with radial kernel (right side) applied

Now, if we perform "majority of voting" on the above results, we conclude "Red" class as a final decision.

(ii) All vs. One SVM Classification

In this type of multi-class SVM classification, each time one of the K classes will be compared with the remaining $(K-1)$ classes [31]. We compare between the Red class (+1 coded) with not-Red class $(-1$ coded), that represents combination of the remaining $(K-1)$, i.e. *Green+Blue* classes. Now if there is one test observation arrive, then we assign the observation to the class for which $\beta_{0k} + \beta_{1k}x_1^* + \beta_{2k}x_2^* + \ldots\ldots + \beta_{mk}x_m^*$ is largest, $\beta_{0k}. \beta_{1k}, \beta_{2k}, \ldots\ldots, \beta_m$ denote the parameters that results from fitting an SVM comparing the Kth class to the others.

$$\text{Example}: \quad \text{SVM}\left(\text{Red}, \text{Not-Red}\right) = \text{Red}, \quad \textit{Not-Red} \geq \textit{Green} + \textit{Blue}$$

Pros
1. SVM works well when the user has no idea of data.
2. It also works well for unstructured data.
3. Kernel trick is the real strength of SVM. Any complex problem or non-separable data issues can be solved by it.

Cons
1. Choosing a suitable kernel is a challenging job.
2. SVM requires long training time for large datasets.
3. It is not easy to fine tune the hyperparameters (like, cost, gamma, etc.) using in SVM.

To load the dataset into our pandas' data frame after it has been imported. Our dataset will now be divided into properties and labels. The first eight columns of the dataset (i.e. attributes) are stored in the X variable, while the labels are stored in the y variable. We'll divide our dataset into training and test divides to minimise overfitting, and this will give us a better picture of how our algorithm fared during the testing phase. Our method is thus tested on unseen data, just as it would be in a real-world application. The script above divides the dataset into two parts: train data and test data. This means that the training set will have 614 records and the test set will have 154 records out of a total of 768 records [32, 33].

For the purpose of simplicity, we've merely demonstrated and explained the KNN classifier's execution. The KNeighborsClassifier from the sklearn.neighbors library must be imported first. Making predictions based on our test results is the final stage. Confusion matrix, precision, recall and f1 score are the most often used metrics for evaluating an algorithm. These metrics can be calculated using the sklearn.metrics confusion matrix and classification report functions. Our KNN algorithm was able to correctly categorise all 30 records in the test set with a 70% accuracy (approx.), which is a decent result. When dealing with high-dimensionality or categorical information, KNN isn't always as effective. The code for all of the other popular classifiers is also shown here [34, 35].

3.8   R Framework

Code No.1: (Using e1071 Package)
In this framework, the input "*iris.csv*" dataset is considered here. It is one of the popular datasets downloaded from UCI ML repository (https://archive.ics.uci.edu/ml/datasets/iris). The data collection has three classes, each with 50 instances, each referring to a different species of iris plant. One class is linearly separable from the other two; however, the subsequent two are not linearly separable. There are four attributes in this dataset, as well as one target class. It's a difficulty with several classifications.

This feedforward multilayer network, which uses the well-known back-propagation architecture, excels in non-linear solutions to ill-defined problems [40]. Each layer in a back-propagation network is fully connected to the layer above it, and there are usually one or two hidden levels. It refers to the delta rule, which calculates the difference between actual and desired outputs. The delta rule error is used to alter the connection weights in most cases. The previous layers are processed until the input layer is reached [36, 39]. The most difficult element of this method is figuring out which input contributed the most to an incorrect result and how to change the input to fix the problem. The back-propagation algorithm iterates over numerous cycles of two processes in its most basic form. An epoch is the name given to each iteration of the algorithm. Because there is no a priori (pre-existing) information in the network, the weights are usually chosen at random before it starts. The algorithm then cycles through the procedures until it reaches a stopping threshold. Gradient descent is a technique, which is involved to decide how much (or whether) a weight should be changed [37, 38].

Pros
1. It is suitable for classification or numeric prediction problems.
2. It is one of the most accurate modelling approaches.

Cons
1. Training process is slow especially if the network topology is complex.
2. Easy to overfit or underfit training data.
3. It is a computationally intensive process.

3.9   Conclusion

The future scope of the work could be analysing the other popular clustering techniques in incremental fashion. The other clustering techniques mean hierarchical clustering, grid-based clustering, constraint-based clustering and so on. In the future, the incremental behaviour of all these clustering algorithms will be analysed. And after a detail analysis, the performances of all clustering algorithms will be compared with each other to provide the fact that which one among them is the most efficient clustering algorithm in the dynamic environment. This analysis could be done by applying various dynamic databases.

3.10   Exercises

Q1. "KNN classifier sometimes referred as a lazy learner whereas Naïve Bayes referred as an eager learner classifier" – Justify this statement.

Q2. How zero-frequency problem can be overcome with the help of "add-1 smoothing technique" in Naïve Bayes algorithm? Explain with a suitable example.

Q3. Suppose you have three classes problems (Red, Green and Blue) to handle with the help of SVM classifier. How can you solve it by using "one vs. one" and "all vs. one" approaches of SVM classifier?

Q4.

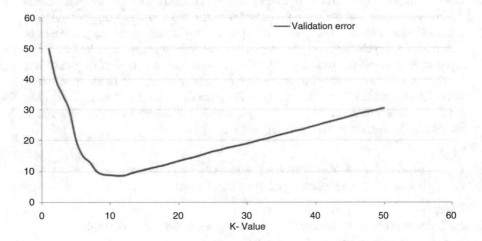

In the image above, which would be the best value for k assuming that the algorithm you are using is k-nearest neighbour? What is the name of this technique of evaluating the initial value of "k"?

Q5.

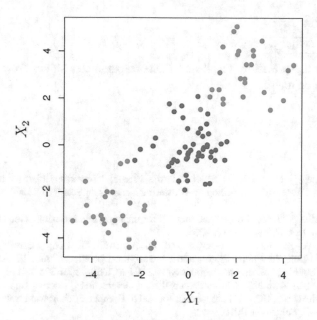

See the above figure where a scatter plot represents a given dataset. For this kind of dataset, which classifier will be suitable for the classification tasks and why?

Q6.

○ ○ △ △ △ △ ○ ○
Non-linearly separable

In the above figure, two kinds of non-linearly separable observations are shown in feature space. How can you enlarge this feature space by using SVM to make it linearly separable and suitable for classification tasks? Explain it with suitable mathematical expressions.

Q7. Outline the benefits of adopting density-based clustering techniques over the partitioning-based clustering.

Q8. Derive the needed expression for Naïve Bayes classifier with the help of Bayes conditional probability theorem.

Q9. What is the difference between "hard margin" and "soft margin" classifier in maximum margin classification technique? Describe with suitable mathematical expressions.

Q10.

TN=97750	FP=150
FN=330	TP=1770

Calculate the precision, recall and F_score parameter values from the above given confusion matrix.

References

1. Eshref Januzaj, Hans-Peter Kriegel, Martin Pfeifle, "Towards Effective and Efficient Distributed Clustering", Workshop on Clustering Large Data Sets (ICDM2003), Melbourne, FL, 2003.
2. S.Jiang, X.Song, "A clustering based method for unsupervised intrusion detections" . Pattern Recognition Letters, PP.802–810, 2006.
3. Guha A., D. Samanta, A. Banerjee and D. Agarwal, "A Deep Learning Model for Information Loss Prevention From Multi-Page Digital Documents," in IEEE Access, vol. 9, pp. 80451-80465, 2021, doi: https://doi.org/10.1109/ACCESS.2021.3084841.
4. A.M.Sowjanya, M.Shashi, "Cluster Feature-Based Incremental Clustering Approach (CFICA) For Numerical Data, IJCSNS International Journal of Computer Science and Network Security, VOL.10 No.9, September 2010.
5. Air-pollution database, WBPCB, URL: 'http://www.wbpcb.gov.in/html/airqualitynxt.php'.
6. Althar, R.R., Samanta, D. The realist approach for evaluation of computational intelligence in software engineering. Innovations Syst Softw Eng 17, 17–27 (2021). https://doi.org/10.1007/s11334-020-00383-2.
7. Anil Kumar Tiwari, Lokesh Kumar Sharma, G. Rama Krishna, " Entropy Weighting Genetic k-Means Algorithm for Subspace Clustering ",International Journal of Computer Applications (0975– 8887),Volume 7– No.7, October 2010.
8. Aristidis Likasa , Nikos Vlassis, Jakob J. Verbeek ," The global k-means clustering algorithm " , the journal of the pattern recognition society, Pattern Recognition36 (2003) 451-461, 2002.
9. B. Naik, M. S. Obaidat, J. Nayak, D. Pelusi, P. Vijayakumar and S. H. Islam, "Intelligent Secure Ecosystem Based on Metaheuristic and Functional Link Neural Network for Edge of Things," in IEEE Transactions on Industrial Informatics, vol. 16, no. 3, pp. 1947–1956, March 2020, doi: https://doi.org/10.1109/TII.2019.2920831.
10. Carlos Ordonez and Edward Omiecinski, "Efficient Disk-Based K-Means Clustering for Relational Databases", IEEE transaction on knowledge and Data Engineering,Vol.16,No.8,August 2004.
11. Carlos Ordonez, "Clustering Binary Data Streams with K-means", San Diego, CA, USA. Copyright 2003, ACM 1- 58113-763-x, DMKD'03, June 13, 2003.
12. CHEN Ning , CHEN An, ZHOU Long-xiang, "An Incremental Grid Density-Based Clustering Algorithm", Journal of Software, Vol.13, No.1,2002.
13. D. Samanta et al., "Cipher Block Chaining Support Vector Machine for Secured Decentralized Cloud Enabled Intelligent IoT Architecture," in IEEE Access, vol. 9, pp. 98013–98025, 2021, doi: https://doi.org/10.1109/ACCESS.2021.3095297.
14. Data Mining concepts and techniques by Jiawei Han and Micheline Kamber, Morgan Kaufmann (publisher) from chapter-7 'cluster analysis', ISBN:978-1-55860-901-3, 2006.
15. Debashis Das Chakladar and Sanjay Chakraborty, EEG Based Emotion Classification using Correlation Based Subset Selection, Biologically Inspired Cognitive Architectures (Cognitive Systems Research), Elsevier, 2018.

16. Dunham, M.H., Data Mining: Introductory And Advanced Topics, New Jersey: Prentice Hall, ISBN-13: 9780130888921. 2003.
17. Govender, P., & Sivakumar, V. (2020). Application of k-means and hierarchical clustering techniques for analysis of air pollution: A review (1980–2019). Atmospheric pollution research, 11(1), 40–56.
18. Guha, A., Samanta, D. Hybrid Approach to Document Anomaly Detection: An Application to Facilitate RPA in Title Insurance. Int. J. Autom. Comput. 18, 55–72 (2021). https://doi.org/10.1007/s11633-020-1247-y
19. H.Witten, Data mining: practical machine learning tools and techniques with Java implementations San-Francisco, California : Morgan Kaufmann,ISBN: 978-0-12-374856-0 2000.
20. Jahwar, A. F., & Abdulazeez, A. M. (2020). Meta-heuristic algorithms for k-means clustering: A review. PalArch's Journal of Archaeology of Egypt/Egyptology, 17(7), 12002–12020.
21. K. Mumtaz, Dr. K. Duraiswamy, "An Analysis on Density Based Clustering of Multi Dimensional Spatial Data", Indian Journal of Computer Science and Engineering, Vol. 1 No 1, pp-8–12, ISSN : 0976-5166.
22. K. Wang et al., "A Trusted Consensus Scheme for Collaborative Learning in the Edge AI Computing Domain," in IEEE Network, vol. 35, no. 1, pp. 204–210, January/February 2021, doi: https://doi.org/10.1109/MNET.011.2000249.
23. Kantardzic, M.Data Mining: concepts, models, method, and algorithms, New Jersey: IEEE press, ISBN: 978-0-471-22852-3, 2003.
24. Kehar Singh, Dimple Malik and Naveen Sharma, "Evolving limitations in K-means algorithm in data Mining and their removal", IJCEM International Journal of Computational Engineering & Management, Vol. 12, April 2011.
25. Khamparia, A, Singh, PK, Rani, P, Samanta, D, Khanna, A, Bhushan, B. An internet of health things-driven deep learning framework for detection and classification of skin cancer using transfer learning. Trans Emerging Tel Tech. 2020;e3963. https://doi.org/10.1002/ett.3963
26. Long, Z. Z., Xu, G., Du, J., Zhu, H., Yan, T., & Yu, Y. F. (2021). Flexible Subspace Clustering: A Joint Feature Selection and K-Means Clustering Framework. Big Data Research, 23, 100170.
27. Lopamudra Dey, Sanjay Chakraborty, Anirban Mukhopadhyay. Machine Learning Techniques for Sequence-based Prediction of Viral-Host Interactions between SARS-CoV-2 and Human Proteins. Biomedical Journal, Elsevier, 2020.
28. Martin Ester, Hans-Peter Kriegel, Jorg Sander, Michael Wimmer, Xiaowei Xu, "Incremental clustering for mining in a data ware housing", 24th VLDB Conference New York, USA, 1998.
29. Michael K. Ng, Mark Junjie Li, Joshua Zhexue Huang, and Zengyou He, " On the Impact of Dissimilarity Measure in k-Modes Clustering Algorithm ", IEEE transaction on pattern analysis and machine intelligence, vol.29, No. 3, March 2007.
30. Naresh kumar Nagwani and Ashok Bhansali, "An Object Oriented Email Clustering Model Using Weighted Similarities between Emails Attributes", International Journal of Research and Reviews in Computer science (IJRRCS), Vol. 1, No. 2, June 2010.
31. Oyelade, O.J, Oladipupo, O. O, Obagbuwa, I. C, "Application of k-means Clustering algorithm for prediction of Students' Academic Performance",(IJCSIS) International Journal of Computer Science and Information security,Vol.7,No. 1, 2010.
32. Rohan Kumar, Rajat Kumar, Pinki Kumar, Vishal Kumar, Sanjay Chakraborty, Prediction of Protein-Protein interaction as Carcinogenic using Deep Learning Techniques, 2nd International Conference on Intelligent Computing, Information and Control Systems (ICICCS), Springer, pp. 461–475, 2021.
33. Sauravjyoti Sarmah, Dhruba K. Bhattacharyya," An Effective Technique for Clustering Incremental Gene Expression data" , IJCSI International Journal of Computer Science Issues, Vol. 7, Issue 3, No 3, May 2010.
34. Steven Young, Itemer Arel, Thomas P. Karnowski,Derek Rose, University of Tennesee, "A Fast and Stable incremental clustering Algorithm", TN 37996, 7th International 2010.
35. Taoying Li and Yan Chen, "Fuzzy K-means Incremental Clustering Based on K-Center and Vector Quantization", Journal of computers, vol. 5, No.11, November 2010.

36. Tapas Kanungo , David M. Mount , "An Efficient k-Means Clustering Algorithm: Analysis and implementation IEEE transaction vol. 24 No. 7, July 2002.
37. Tavallali, P., Tavallali, P., & Singhal, M. (2021). K-means tree: an optimal clustering tree for unsupervised learning. The Journal of Supercomputing, 77(5), 5239–5266.
38. Weka, Waikato environment for knowledge environment - http://www.cs.waikato.ac.nz/ml/weka/.
39. Xiaoke Su, Yang Lan, Renxia Wan, and Yuming, "A Fast Incremental Clustering Algorithm ", international Symposium on Information Processing (ISIP'09), Huangshan, P.R.China, August-21–23,pp:175–178,2009.
40. Zuriana Abu Bakar, Mustafa Mat Deris and Arifah Che Alhadi, "Performance analysis of partitional and incremental clustering", SNATI, ISBN-979-756-061-6, 2005.

Chapter 4
Data Classification and Incremental Clustering Using Unsupervised Learning

4.1 Introduction

Clustering is a crucial aspect of data mining. As the amount of data expands and the processing capacity of computers increases, clustering becomes more important. Artificial intelligence, pattern recognition, economics, ecology, psychiatry, marketing, biology, and machine learning all use clustering applications to some extent. Clustering is the process of organizing data into classes or clusters. The objects within a cluster are similar but quite different from those in other clusters [1]. Clustering is viewed through data modeling, founded in mathematics, statistics, and numerical analysis. Clusters refer to hidden patterns in machine learning, the search for clusters is unsupervised learning, and the resulting system is a data notion. Clustering is thus the unsupervised acquisition of a hidden data idea. Data mining deals with enormous databases, which adds to the computational demands of clustering analysis. As a result of these difficulties, powerful and widely applicable data mining clustering approaches have emerged. In certain applications, clustering is referred to as data segmentation since it divides big datasets into categories based on their resemblance. Outliers (values that are "far away" from any cluster) may be more interesting than usual examples; hence clustering can be utilized for outlier detection [2, 3]. The detection of credit card fraud and monitoring illegal activity in electronic commerce are two examples of outlier detection applications [4].

4.2 Literature Review

The researchers suggested a global K-means algorithm. The most well-known clustering algorithm is K-means clustering, which can reduce clustering error to a minimum. In terms of clustering error, the K-means algorithm finds locally optimal

S. Chakraborty et al., *Data Classification and Incremental Clustering in Data Mining and Machine Learning*, EAI/Springer Innovations in Communication and Computing, https://doi.org/10.1007/978-3-030-93088-2_4

solutions. It's a quick iterative approach used in many clustering software. Although it is a local process, K-means has a significant drawback in that its performance is highly dependent on the initial beginning conditions. As a result, the cluster center must be declared first. As a result, this technology has not acquired widespread acceptance, and its clustering mechanism causes problems in many practical applications [5, 6]. As a result, numerous runs with different initial placements of the cluster centers must be scheduled to find near-optimal solutions using the K-means method. A global K-means algorithm is proposed to handle this problem, which is a deterministic global optimization approach that does not require any initial parameter values and uses the K-means algorithm as a local search strategy. Rather than selecting initial values for all cluster centers at random, as most global clustering algorithms do, the suggested technique works in stages, seeking to add one new cluster centre ideally at each stage. To use this methodology with M clusters, start with one cluster (K = 1), and locate its ideal position, which corresponds to the dataset's centroid. When this strategy is used on two clusters (K = 2), N K-means algorithm executions are examined. When K = 1, the first cluster center is always in the best position, but when K = 2, the second cluster center is in the position of the data point (X_n = 1, 2, ..., N). This study invents the global K-means algorithm and proposes some variations based on the main notion of the method [7, 8]. Figure 4.1 shows basic stages in clustering.

The suggested technique is deterministic, does not rely on any starting cluster centre positions and does not contain any empirically adjustable parameters; hence it solves all of the issues that plague the K-means algorithm and its stochastic expansions. Figure 4.2 projects performance results for the Iris dataset.

K-means clustering is very simple to use, but it creates lots of problems with high-dimensional data and especially with relational databases [9]. An innovative concept of disk-based K-means algorithm was developed by Carlos Ordonez and Edward Omiecinski [2]. This algorithm can perform efficient disk-based K-means clustering in relational databases. In many applications, data may be stored in files or tables format. Many dimension values may be zero for many points in very high-dimensional data [10]. As a result, a large amount of storage space for that file or

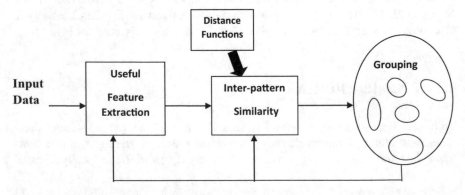

Fig. 4.1 Basic stages in clustering

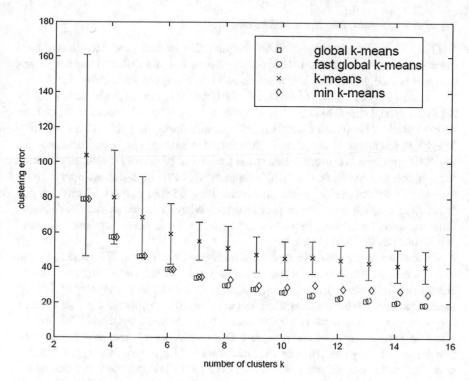

Fig. 4.2 Performance results for the Iris dataset

table is squandered, and it also necessitates a large amount of CPU calculation, slowing down the clustering procedure. The storage spaces of memory are also wasted due to the storage of data matrices. Most of the values of these data matrices are filled with null entries. That's why an innovative concept of disk-based K-means algorithm is developed that can perform efficient disk-based K-means clustering in relational databases [11]. It can cluster high-dimensional large datasets in RDBMS. It can perform heavy disk input-output (I/O), and its memory requirements are low. This disk-based K-means algorithm is also referred to as RKM (relational K-means). The main motivation of this RKM algorithm is divided into two parts; they are as follows:

(i) *Algorithmic improvements*

The initialization of centroids in algorithmic improvements is based on the dataset's global mean and covariance. To achieve faster convergence, sufficient statistics are paired with periodic M steps. To increase the quality of the answer, the algorithm employs cluster splitting. Because it has particular operations for sparse matrices, the technique can successfully handle transaction data. In general, each procedure run only necessitates three scans of the dataset, plus a single additional run to compute the global mean and covariance.

(ii) *Disk organisation of input dataset and matrices*

Disc I/O is minimized by using an adequate disc organization and the K-means method matrix access pattern. The memory needs are minimal, and memory management is straightforward. The input dataset can be organised one row per dimension value per point in one of two ways: first, matrices can be organised as binary files with a fixed structure during the algorithm execution, and, second, matrices can be organised as binary files with a fixed structure during the algorithm execution. The RKM technique is tested on both numeric generated and actual datasets [12, 13]. This approach is also compared to the traditional K-means technique. The main focus of this research is on the RKM's capacity to cluster high-dimensional transaction files. The number of points, dimensionality, desired number of clusters and transaction size all have linear performance. When transaction size is constant, dimensionality has a minor impact on performance. There are no fundamental memory limitations in RKM [14].

Carlos Ordonez proposed a novel clustering approach using K-means to cluster binary datasets. For a variety of reasons, binary datasets are fascinating and helpful. Binary data refers to information that is easily accessible on a computer and can be used to express category information. Because there is no noise in the data, clustering on binary data streams is a helpful and straightforward approach. Data can be effectively stored, indexed, and retrieved. This method uses K-means to improve the clustering of binary data. Simple adequate statistics for binary data, quick distance computation for sparse binary vectors, and sparse matrix operations are among the enhancements [15]. It also explains how to incorporate all these modifications into other K-means algorithms. There are two methods for clustering binary data streams:

(i) *Sparse matrix operations and simple sufficient statistics*

K-means clustering is used on sparse distance calculation and simpler adequate statistics to speed up clustering on binary data streams. The distances between the null transaction and all centroids in sparse distance calculation are pre-computed [16]. Only differences for non-null dimensions of each transaction are computed as transactions are read to establish cluster membership. This computation boosts performance significantly but does not effect on the correctness of the results.

(ii) *K-means variants for binary data streams*

Several K-means versions are used to cluster binary data streams and assess their performance effectively. Online K-means, incremental K-means, scalable K-means, and standard K-means are the K-means versions. Incremental K-means and scalable K-means were the best quality solutions based on their experiments on real datasets. However, online and incremental K-means were the fastest in terms of performance. Standard K-means is the slowest of them all [17]. Figure 4.3 describes how the online K-means, standard K-means, incremental K-means and scalable K-means vary their performances with time while using transaction datasets.

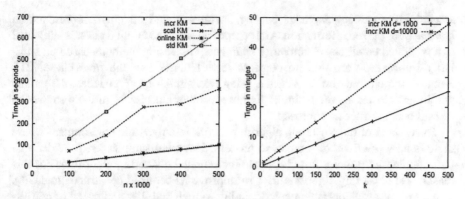

Fig. 4.3 Performance varying of K-means variants on transaction datasets

They developed an approach based on clustering of categorical data using the K-modes algorithm [18]. Clustering categorical datasets have become a major problem in recent years. The main purpose of this method is to use an optimization framework to construct the update formula for the K-modes clustering algorithm with the new dissimilarity measure and the algorithm's convergence. The new dissimilarity measure's object cluster membership assignment method and mode update formulas are also rigorously shown to minimize the objective function in this study. According to testing results, the K-modes algorithm with the new dissimilarity measure performs better in clustering accuracy than the old K-modes approach [19].

Aweighted email attribute similarity-based data mining technique is presented for email clustering to locate email groups. Email attribute means email inbox, outbox, subjects, attachments, etc. This chapter proposes a concept that provides the idea of clusters of similar email attributes based on their weighted matching. Specific user-defined weights are sometimes assigned to the similarity measured between a pair of email attributes to calculate the similarity between pairs of emails. This chapter uses well-known K-means clustering technique, and emails are collected from the "Enron email corpus dataset". Most of the emails are text-based email, that's why this chapter uses various string similarity algorithms present today [20]. Such as

(a) Dice similarity
(b) Cosine similarity
(c) Bleu similarity
(d) TF-IDF similarity
(e) Jaccard similarity

Based on all these string similarity algorithms, pairs of attributes, subject, message, from-mail-address, and so on are clustered properly. "Figure 4.3" drawn below can graphically represent the main concept of this chapter.

First of all, emails are collected from the Enron email corpus dataset and stored as a primary data source. After that preprocessing is applied to that primary data source. Preprocessing includes parsing, stemming and an email representation approach for parsed information. After preprocessing, the email dataset is subjected to a weighted email object similarity approach. Clustering methods based on similarity information are used to construct email clusters in the final phase [21]. Because the clusters are constructed using text similarity approaches, the emails within one cluster are often similar to those in other clusters. Figure 4.4 shows the overall email clustering approach.

Today most of the clustering algorithms work in a dynamic environment where there is no a priori set of samples to process. However, samples are provided one iteration and proposed a fast and stable incremental clustering algorithm. For this reason, the clustering algorithm tries to improve its centroid constructs gradually. Thus, several problems are emerged relating to their stability and speed of convergence. This fast and stable incremental clustering algorithm is computationally modest and imposes minimal requirements. This approach is useful in applications where fast, online clustering of large databases is required [22]. Generally, reaching stability in a dynamic environment means centroid is changing with minimal tolerance. This chapter uses a learning rate, which decays with the time that helps it to reach proper stability. To perform this work, this chapter introduces a proposed framework for efficient incremental clustering using a winner-take-all (WTA) inspired partition approach.

Fig. 4.4 The overall email clustering approach

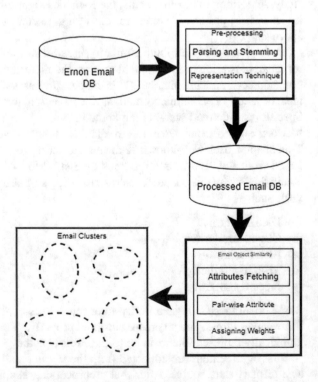

The proposed clustering algorithm employs the WTA competitive learning but with an online aspect where it continuously updates centroids based on the input data stream or observations. This chapter presents a fast and scalable incremental clustering algorithm that is modest in computational and memory resources. In particular, starvation of prototypes and the general stability of the process are guaranteed using specific statistical metrics given the input data has a reasonably stable structure [23].

The above figure can be easily detected the points of main clusters and the noise points. The density in the noise areas is lower than in any of the clusters, and each cluster has a mean point density that is significantly higher than the density outside of the cluster. Figure 4.5 express proposed clustering framework [24].

This chapter also uses the "lemmata" technique for validating the correctness of this clustering algorithm. It also discusses "Eps-neighborhood of a point and minimum number (*MinPts*) of points". It also evaluates the performance of DBSCAN and compares it with the performance of CLARANS because this is the first and only clustering algorithm designed for the purpose of KDD. The results of these tests show that DBSCAN is substantially more effective than the well-known method CLARANS in discovering clusters of any shape. Furthermore, DBSCAN exceeds CLARANS in terms of efficiency by a factor of at least 100, according to the trials. Figure 4.6 shows density-based notion of clusters.

Many clustering techniques exist in the literature, but this book mainly focuses on the K-means and DBSCAN clustering algorithms. The above chapters describe how these algorithms are introduced and how their characteristics change when they

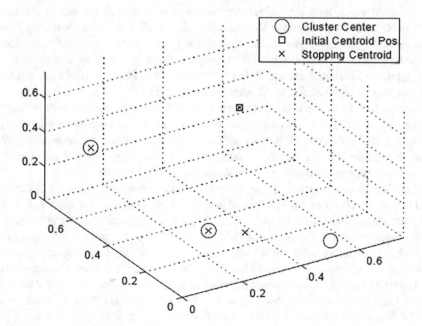

Fig. 4.5 Proposed clustering framework

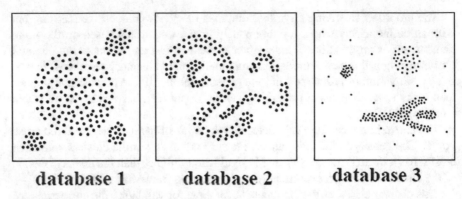

database 1   **database 2**   **database 3**

Fig. 4.6 Density-based notion of clusters

are applied to large datasets. The above chapters also define their performances on various datasets. They also explain the limitations of these algorithms and the different ways to overcome those limitations. But all of these algorithms are not efficient with dynamic databases [25].

The researchers discussed a brief overview of K-means clustering summarised well-known clustering techniques and their wideness of applications. This work also discusses all kinds of challenges and issues in designing those clustering methods and also leads towards some emerging and significant research directions. After a rigorous analysis of all competitive clustering algorithms, it concludes that no clustering method is the best and dominates other algorithms across all application domains. Machine learning and pattern recognition in research communities indicate several issues to improve the overall understanding of data clustering. This chapter briefly discusses the issues, such as the ground truth data should have both static and dynamic data along with the popularity of quantitative and qualitative attributes, and it should include datasets from various domains. However, most of the popular datasets available in some famous online repositories are practically small and limited to static and less diverse data. A good clustering method should create stable, consistent data partitions and provide stable solutions. It is always advisable to choose computationally efficient optimal solutions for clustering problems, especially for a large scale of data, as finding the global optimal solutions for some basic clustering algorithms (like K-means) is NP-hard [26].

The developers proposed an entropy-weighted power-based K-means clustering method that overcomes the problem of poor local minima, and it is highly scalable for high-dimensional data as well. Through annealing, local minima are avoided here. Besides that, the proposed objective function is minimized by using majorization-minimization (MM) algorithm. This presented an approach to use trees for unsupervised learning tasks. All existing clustering trees algorithms are suboptimal. Therefore, this approach helps to learn the optimal clustering trees in linear time space. This proposed K-means tree gives the centroids of clusters as output [27, 28]. It can learn the cluster centroids more efficiently in less query time.

It leads to the problem of learning centroids and tree parameters jointly, and these have been solved using the quadratic penalty method.

The researchers introduced an extended K-means data clustering algorithm with an additional cluster to hold the outliers simultaneously. In this algorithm, two parameters are used internally, and one objective function is also used to maintain the efficiency and effectiveness of the algorithm. The first parameter is the maximum number of outliers produced regardless the value of the second parameter value. A larger value of the second parameter leads to less number of outliers. This idea proposed a flexible subspace clustering that merges feature selection and K-means in a unified framework to improve the overall performance of clustering. To make this approach more robust, an $l_{2,p}$ norm is associated with the objective function. This "p" varies on different data and increases robustness (robust to noise and redundant features of big data) [29]. This proposed an optimal unsupervised K-means clustering algorithm without initialising any parameters but obtaining an optimal number of clusters as a result. In this algorithm, an entropy penalty term for adjusting bias is considered initially and then starts a learning procedure to find the optimal number of clusters. It uses the number of points as the initial number of clusters and discards extra clusters during iterations, and finally the optimal number of clusters can be automatically found according to the structure of data. I we devised a new locally operated private K-means clustering algorithm for the Euclidean K-means problem. The additive error was lowered by this differentially private K-means technique, while the multiplicative error remained same. In terms of its reliance on database size n, the received additive error is ideal [30].

This devised an improved version of Lloyd's K-means clustering algorithm that is enriched with speed and simplicity and suitable for high-dimensional data without initializing many clusters. Power K-means embeds the K-means problem in a continuum of better-behaved problems. This algorithm guarantees to reduce the K-means objective at each step with the help of memorization-minimization algorithm. Its complexity is similar with the Lloyd's K-means complexity. This method is generally an extension of k-harmonic means clustering, overcoming its limitations to low-dimensional problems. It introduced an accelerated version of a hybrid hierarchical density-based clustering (HDBSCAN). It supports variable density clusters and eliminates the need for the difficult-to-tune distance scale parameter ϵ. It has a comparative asymptotic performance to DBSCAN. It proposed a modified BLOCK-DBSCAN approach for large-scale data and also used to overcome the redundant distance computations problem of grid technique-based fast DBSCAN useless in high-dimensional data space [31]. This approach is divided into some sub-techniques: first is to find more core points at one time rather than grid technique, and it uses $\epsilon/2$ norm ball to discover inner core blocks, second is to devise a fast algorithm to check whether two inner core blocks are density reachable or not and third is to build a cover tree to accelerate the process of density computations. BLOCK-DBSCAN has two types, L_2 for relatively high-dimensional data and L_∞ for high-dimensional data. It uses a cover tree to exclusively find the inner core blocks, border points and outer core points. Now a fast approximate algorithm merges each pair of inner core blocks into the same cluster if they are density reachable from each other. Each outer core point is also merged after satisfying density reachability criteria. At the final stage, each border point will be assigned to corresponding cluster by the BLOCK-DBSACN.

The researchers discussed a survey on combining clustering techniques for air pollution studies. This chapter reviews various research works, such as links to meteorological conditions, etc. The K-means clustering technique and average linkage-based hierarchical techniques are mostly used in air pollution studies. It proposed a three-way clustering on an improved DBSCAN algorithm (3W-DBSCAN) for better understanding the relationship between a data point and a cluster. Two nested sets lower and upper bounds represent a cluster, and these two bounds classify a data object into three statuses, belong to, not belong to, and ambiguity. Lower-bound objects directly belong to the clusters. Ambiguous objects in the upper bound are in a boundary region and may belong to any cluster. Objects beyond the upper bound do not exist in the same cluster. This representation is matched with human cognitive thinking and produces better results [32].

4.3   Types of Clustering

The major clustering methods can be classified into various numbers of categories. "Figure 4.7" shows those different categories of clustering [4].

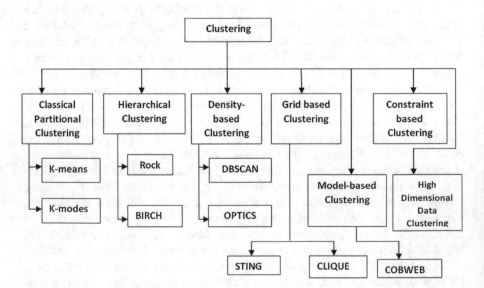

Fig. 4.7 Hierarchy of general clustering methods

4.4   Popular Clustering Techniques

The K-means algorithm is the most popular representative-based clustering algorithm. It is greedy and computationally efficient. Below is the pseudocode for the K-means algorithm. It indicates how the K-means clustering algorithm is applied to cluster a set of N data vectors (X). It describes that after initialisation of K cluster centroids, K-means is applied on the X dataset to form some clusters (C). The algorithm simply computes the minimum distance between cluster centroids and the data vectors using different distance measure functions discussed above to form the clusters. And it repeats these steps until all the data vectors are properly clustered.

4.4.1   Flowchart of K-Means Clustering

The flowchart of K-means clustering shows an overall schematic diagram regarding this clustering technique. It describes the same concept as the K-means algorithm describes earlier. "Figure 4.8" depicts the pictorial representation of the K-means clustering algorithm. Normally K-means works according to the four basic steps described below. A detailed hands-on illustration of the K-means clustering algorithm with example.

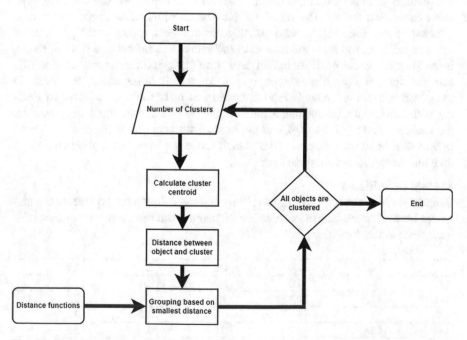

Fig. 4.8 Flow chart depicting the K-means clustering algorithm

4.4.2   Applications of K-Means Clustering

K-means clustering algorithm is widely used in educational research, market research, medicine and biomedical, neuroscience, software evolution, social network analysis, image segmentation, data mining, climatology, a grouping of shopping items, fraud detection, etc. Today, a system is given to assess student results using cluster analysis and basic statistical techniques to organize their scores according to their performance level. The K-means clustering approach is widely employed in this circumstance [33].

K-modes clustering is an extension of K-means clustering, which is more suitable for efficiently computing the clusters of categorical data. It replaces the initial assumptions with modes instead of means and also uses methods. To overcome the challenges of strict boundaries between clusters EM (expectation-maximization) clustering technique is adopted where the new means are calculated by the weighted measures of each cluster object's probability of membership. Besides that, some other K-means variations have been introduced, like, online K-means clustering, power K-means clustering, entropy K-means, etc.

4.4.3   K-Medoids Clustering

The outliers or noisy data influence the K-means clustering where extremely large value can distort the distribution of the data. As all of us know, means are always affected by the large values, whereas medoids are more robust in the presence of noise or outliers, and the outliers or extreme values less influence a medoid than a mean [34]. For example, if we have a data point (1, 2, 300), the mean value is 101, and the formed cluster prototype is not an actual observation. However, in the k-medoids method, medoid is an extended version of the median. If we are dealing with one feature, the median is the middlemost value if arranged. For example, the median value of (1,4,8) is 4, and the median value of (1,4,6,10) can be thought to be 4 or 6 or an average of 4 and 6, which is 5. When you are dealing with more than one feature, it is called medoids.

Medoid Definition-I

Suppose there are a total five observations and the below table represents the distances between each pair of observations. Then choose the maximum distance for each observation row-wise.

	O1	O2	O3	O4	O5
O1	0	1.5	2	2.4	**3.2**
O2	1.5	0	4	**4.1**	3.1
O3	2	**4**	0	2.2	2.1
O4	2.4	4.1	2.2	0	**5.9**
O5	3.2	3.1	2.1	**5.9**	0

	Max distance
O1	**3.2 (min)**
O2	4.1
O3	4
O4	5.9
O5	5.9

From the above table, we choose the minimum cost (distance) as a final medoid.

Medoid Definition-II

This method is divided into three steps:

Step-1: Calculate the mean.
Step-2: Find distances of all the points from the mean.
Step-3: The one with the lowest distance from the mean is the medoid.

	Distance from mean
O1	3.8
O2	**2.5 (min)**
O3	4
O4	5.5
O5	5.6

This second medoid is more preferable to follow.

However, there are four cases that are considered to determine a good replacement strategy between O_j and O_{rand}. Suppose p is considered as non-representative object. Figure 4.9 shows different cases of the cost function in k-medoids clustering

Illustration (Swap Step and Cost)

Suppose the following table represents a set of 2D points and their corresponding cost in terms of distance.

X	Y	$M_1(1,5)$	$M_2(3,3)$	Cost
1	5	0	4	0
2	4	2	2	2
2	7	3	5	3
3	5	2	2	2
1	4	1	3	1
4	2	6	2	2
1	6	1	5	1
1	9	4	8	4
2	4	2	2	2
4	4	4	2	2
3	3	4	0	0
5	4	5	3	3
Total cost				22

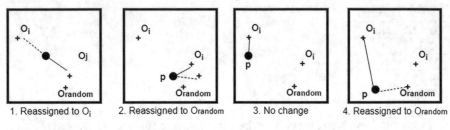

1. Reassigned to O_i 2. Reassigned to Orandom   3. No change   4. Reassigned to Orandom

● data object
+ cluster center
— before swapping
---- after swapping

Fig. 4.9 Different cases of the cost function in k-medoids clustering

We will assume $M_1(1,5)$ and $M_2(3,3)$ as the two initial cluster representative objects. The individual cost can be calculated as a subtraction between the distances of two points. The total sum of cost is 22 in the first case. In the second case, we replace the $M_1(1,5)$ representative object with the new object $M_1(1,9)$ and calculate the new total cost again. The below table represents case 2.

X	Y	$M_1(1,9)$	$M_2(3,3)$	Cost
1	5	4	4	4
2	4	6	2	2
2	7	3	5	3
3	5	6	2	2
1	4	5	3	3
4	2	10	2	2
1	6	3	5	3
1	9	0	8	0
2	4	6	2	2
4	4	8	2	2
3	3	8	0	0
5	4	9	3	3
Total Cost				26

Therefore, the total cost is increased. So, we should discard this replacement or swap the move. Instead, we swap the second representative object $M_2(3,3)$ by the object (4,4) keeping the first object intact and calculate the total cost.

X	Y	$M_1(1,5)$	$M_2(4,4)$	Cost
1	5	0	4	0
2	4	2	2	2
2	7	3	5	3
3	5	2	2	2
1	4	1	3	1
4	2	6	2	2
1	6	1	5	1
1	9	4	8	4
2	4	2	2	2
4	4	4	0	0
3	3	4	2	2
5	4	5	1	1
Total cost				20

Therefore, the total is 20 that is less than the total cost of the first case. So, we should consider this replacement or swap step. This is an iterative process, and we should move for all the objects in a similar way [35, 36]. The computational complexity of k-medoids of each iteration is $O(k(n-k)^2)$. This method becomes very costly if the value of n and k increases. Like K-means clustering, it also has the same limitation on the specification of the initial number of clusters (k). Figure 4.10 projects the clustering of a set of objects based on the DBSCAN.

4.5   Flowchart of DBSCAN Clustering Algorithm

The flowchart of DBSCAN algorithm is shown below.

"Figure 4.11" represents the flow chart of DBSCAN clustering algorithm. First it assumes the minimum number of points (Minpts), and the minimum radius of each cluster is represented as ε (eps), and the total number of data objects are represented as D. Now, all visited points which are ε-neighborhood are marked as P in the dataset D. Now, those points are assigned as N, and on N two conditions are applied that means the size of N (total number of ε-neighborhood points) must be greater than Minpts and cluster radius must be greater than the ε. If N is greater than Minpts, then those points are included into clusters and expand clusters, or else P is marked as noise or outlier [37].

However, the height of the blocks represents the distance between the clusters. It can be noted from Fig. 4.12 that 17 and 26 are close to each other. Therefore, they formed the first cluster (c_1). After that, they merged with 28 (c_2). Then 4, 6, and 10 are combined into another cluster (c_3), followed by 3, which is merged with it (c_4). Similarly, this iterative process is going on until merging all the points and forming a final cluster after merging 11 with them.

There are several methods to calculate the distance between the clusters to decide the rules for hierarchical clustering. Among these, all linkage methods,

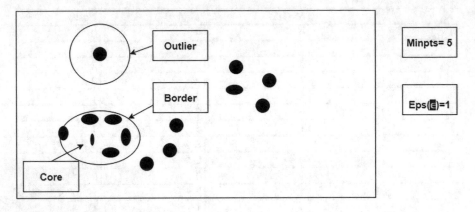

Fig. 4.10 Clustering of a set of objects based on the DBSCAN

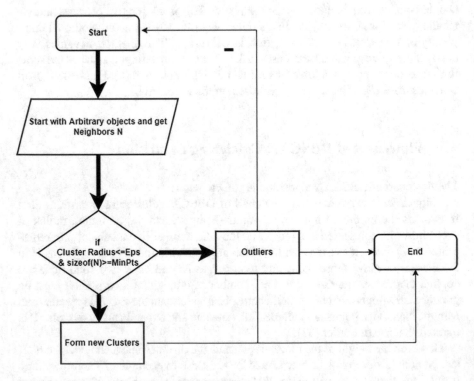

Fig. 4.11 Flow chart depicting the DBSCAN clustering algorithm

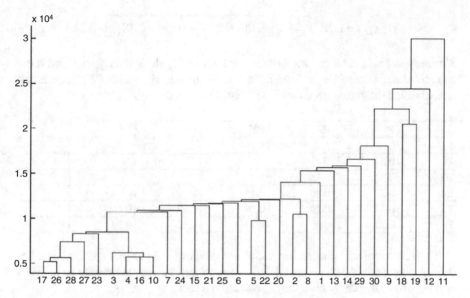

Fig. 4.12 Dendrograms of hierarchical clustering of bladder cancer dataset

complete and mean linkages are the most widely used methods. However, the computation of the mean vector for categorical data can be difficult [38].

4.6   Illustrations

Suppose we have a following example dataset:

Observations	x	y
P1	0.40	0.53
P2	0.22	0.38
P3	0.35	0.32
P4	0.26	0.19
P5	0.08	0.41
P6	0.45	0.30

A. **Agglomerative Clustering (Bottom-up Strategy)**

First, apply agglomerative clustering on the above demo dataset to do the clustering. We consider the Euclidean distance measure to calculate the distance between two points and use single-linkage method. For example, we can calculate the distance between P_1 and P_2 in the following way:

$$D(x,y) = D(P_1,P_2) = \sqrt{(0.40-0.22)^2 + (0.53-0.38)^2} = 0.23$$

Similarly, we will calculate the distance among each pair of data points, and it can be represented in the form of the distance matrix below. This kind of distance matrix is also called the proximity matrix in hierarchical clustering.

	P1	P2	P3	P4	P5	P6
P1	0					
P2	0.23	0				
P3	0.22	0.15	0			
P4	0.37	0.20	0.15	0		
P5	0.34	0.14	0.28	0.29	0	
P6	0.23	0.25	**0.11**	0.22	0.39	0

Therefore, we will choose the minimum distance (single-linkage method), i.e., 0.11, that maps two data points (P_3,P_6). So, we build the first subcluster in the form of a dendrogram tree by grouping P_3 and P_6 together.

Now, the updated proximity matrix is represented as

	P1	P2	P3,P6	P4	P5
P1	0				
P2	0.23	0			
P3,P6	0.22	0.15	0		
P4	0.37	0.20	0.15	0	
P5	0.34	**0.14**	0.28	0.29	0

Therefore, the next minimum distance is 0.14, which maps two data points (P_2,P_5). So, we build the second subcluster in the form of dendrogram tree by grouping P_2 and P_5 together.

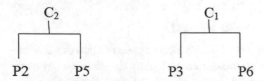

To update the distance matrix:

1. $\min(\text{dist}(P_3,P_6),P_1) = \min(0.22,0.23) = 0.22$
2. $\min(\text{dist}(P_3,P_6),(P_2,P_5)) = \min(0.25,0.15) = \textbf{0.15 (min)}$
3. $\min(\text{dist}(P_3,P_6),P_4) = 0.15$
4. $\min(\text{dist}(P_3,P_6),P_5) = 0.28$

(P_2,P_5) and (P_3,P_6) are mapped into a minimum value, so they are grouped together. Now, the updated proximity matrix is represented as

	P1	P2,P5,P3,P6	P4
P1	0		
P2,P5,P3,P6	0.22	0	
P4	0.37	**0.15**	0

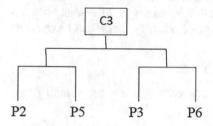

Therefore, the next minimum distance is 0.15 that maps two data points (P_2,P_5,P_3,P_6) and P_4. So, they are grouped together.

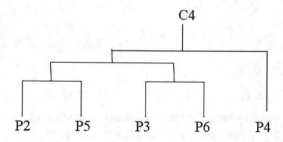

Now, the updated proximity matrix is represented as

	P1	P2,P5,P3,P6,P4
P1	0	
P2,P5,P3,P6,P4	0.22	0

So, finally, P1 is merged with the (P_2,P_5,P_3,P_6,P_4) point with respect to the value of 0.22. So, the final dendrogram of this hierarchical agglomerative clustering is

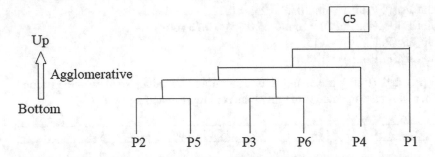

Agglomerative clustering is the most popular hierarchical clustering technique with a wide variety of applications and follows greedy approach while joining. It can be used to identify fake news, analyze documents, cluster US senators into their respective parties from Twitter data, analyze DNA sequences of animals to construct the phylogenetic tree and also track viruses, identify fraudulent or criminal activity, etc. [39].

B. Divisive Clustering (Top-down Strategy)

We are going to apply the divisive clustering technique for the similar demo dataset described below again:

Observations	x	y
P1	0.40	0.53
P2	0.22	0.38
P3	0.35	0.32
P4	0.26	0.19
P5	0.08	0.41
P6	0.45	0.30

Step-1

Similar to the agglomerative method, we consider the Euclidean distance measure for calculating distance between two points and use the complete-linkage method (opposite of the single-linkage method to receive a similar result, as it is a top-down approach). For example, we can calculate the distance between P_1 and P_2 in the following way:

$$D(x,y) = D(P_1, P_2) = \sqrt{(0.40 - 0.22)^2 + (0.53 - 0.38)^2} = 0.23$$

Similarly, we will calculate the distance among each pair of data points, and it can be represented in the form of a distance matrix below. This kind of distance matrix is also called the proximity matrix in hierarchical clustering.

	P1	P2	P3	P4	P5	P6
P1	0					

P2	0.23	0				
P3	0.22	0.15	0			
P4	0.37	0.20	0.15	0		
P5	0.34	0.14	0.28	0.29	0	
P6	0.23	0.25	0.11	0.22	0.39	0

Step-2

Now, we compute a minimum spanning tree (MST) from the above proximity matrix by using either Kruskal's or Prim's algorithm. We are using Prim's algorithm to arrange all the distances in ascending order.

Edge	Cost
P3,P6	0.11
P2,P5	0.14
P2,P3	0.15
P3,P4	0.15
P2,P4	0.20
P1,P3	0.22
P4,P6	0.22
P1,P2	0.23
P1,P6	0.23
P2,P6	0.25
P3,P5	0.28
P4,P5	0.29
P1,P5	0.34
P1,P4	0.37
P5,P6	0.39

Now, the final MST can be constructed by connecting all the edges between points (starting from the minimum) until all the points will participate and no closed-loop or circuit will be formed. So, following the greedy approach, we first connect P_3 and P_6 with the lowest distance (0.11), then connect P_2 and P_5 with the second-lowest distance 0.14, then connect P_2 with P_3 with the third-lowest distance 0.15, and so on.

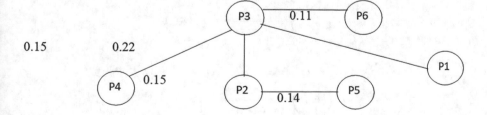

Step-3 Therefore, we gradually apply the complete-linkage method to break the edges of the final MST according to the maximum cost or distance, and we are able to create new clusters by considering those largest distances.

(i) First, we break the edge whose cost is 0.22, i.e., (P_1, P_3). So, two clusters get formed cluster 1 consists of P_1 and cluster 2 consists of $(P_2, P_3, P_4, P_5, P_6)$.

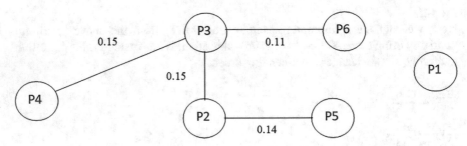

(ii) Now, we will seek for the next maximum-distance greedily, and we will choose (P_3, P_4) as the next highest cost, i.e., 0.15, and break that edge to create three separate clusters, cluster 1={P_1}, cluster 2={P_2, P_3, P_5, P_6} and cluster 3={P_4}. We will continue this splitting iteration until each new cluster contains only a single object.

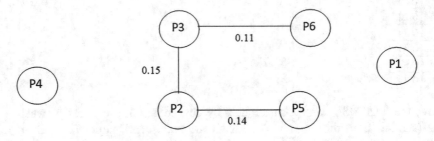

(iii) Now, the next maximum cost is (P_2, P_3), i.e., 0.15, and we will break that edge. So, our final four clusters will be cluster 1={P_1}, cluster 2={P_2, P_5}, cluster 3={P_4} and cluster 4={P_3, P_6}.

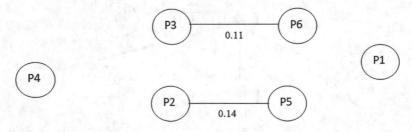

(iv) Then the next maximum distance or cost is 0.14, i.e., (P_2,P_5), and we will break that edge. So, our final five clusters will be cluster 1={P_1}, cluster 2={P_2}, cluster 3={P_4}, cluster 4={P_3,P_6} and cluster 5={P_5}.

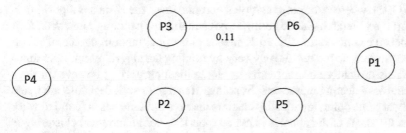

(v) At the final step, we will break the last edge (P_3,P_6) whose cost is 0.11 and separate all the data points into individual clusters.

So, our final six clusters will be cluster 1={P_1}, cluster 2={P_2}, cluster 3={P_4}, cluster 4={P_3}, cluster 5={P_5} and cluster 6={P_6}.

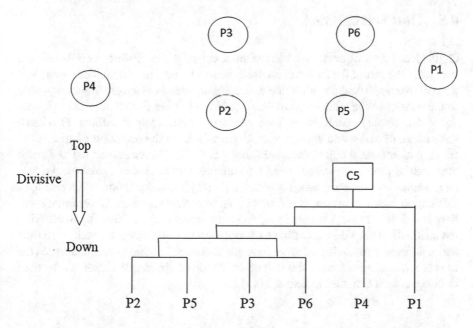

So, if we look and evaluate this final dendrogram (received from agglomerative clustering) from top to bottom direction, then we will just receive the divisive process that we have demonstrated recently.

4.7   Hierarchical vs. K-Means Clustering

The computational complexity of hierarchical clustering is $O(n^2)$, and space complexity is $O(n^3)$, whereas the computational complexity of the K-means algorithm is $O(n)$. Therefore, hierarchical clustering is not suitable for handling large data, as it is too expensive computationally. In K-means clustering, random choice of initial clusters is required, so the results may vary by running the algorithm multiple times [40, 41]. This difficulty does not exist in hierarchical clustering; however, it does have difficulties selecting merge or split points. If merge or split decisions are made incorrectly at some point, low-quality clusters may result. K-means is found to work well when the shape of the clusters is hyper-spherical, but hierarchical clustering is suitable to handle the arbitrary shape of clusters. Some extended versions of hierarchical clustering technique are introduced for improving the quality of clustering. For example, BIRCH, ROCK, and Chameleon algorithms introduce multiple-phase clustering.

4.8   Outlier Analysis

Outliers are data objects that behave in a substantially distinct or inconsistent way from the rest of the data and do not conform to the data model. For example, a general manager's wage at an office will naturally stand out as an outlier compared to the salaries of the other employees. Outlier mining is a term that refers to a variety of data mining operations that are used to find and analyze outliers. This has a wide range of real-world applications. For example, in the detection of credit card fraud, outliers could signal fraudulent behavior [42]. We examined the statistical approach of detecting outliers using interquartile range. Outlier detection can also be classified as a distance-based strategy, a density-based local outlier approach, or a deviation-based approach, in addition to the statistical approach. Determining outliers is a difficult task. Outliers in regression models may be found by identifying the residuals to get a decent estimate of data extremeness; however, outliers in time series data are difficult to find since they are disguised in trend or seasonality. In the case of a multidimensional dataset, a combination of dimension values can be used to discover data extremes as outliers [43].

4.9   Conclusion

Unsupervised learning, i.e., clustering, is one of the effective methods for knowledge discovery and data mining applications. This chapter describes the fundamental concepts of clustering and some state-of-the-art clustering techniques (partitioning, density-based and hierarchical) that have many applications in real-world data where expert advice is not present. Clustering always concerned about the groups, even if there is no group structure. In this chapter, various forms of clustering similarity criteria have been discussed. After the clusters have been constructed, the assessment criteria are summed up to determine how well the clusters perform, how valid, and how good they are. This chapter provides a comprehensive comparison of clustering algorithms and some of their specific uses.

References

1. Aristidis Likasa , Nikos Vlassis, Jakob J. Verbeek ," The global k-means clustering algorithm " , the journal of the pattern recognition society, Pattern Recognition36 (2003) 451-461, 2002.
2. Carlos Ordonez, "Clustering Binary Data Streams with K-means", San Diego, CA, USA. Copyright 2003, ACM 1- 58113-763-x, DMKD'03, June 13, 2003.
3. K. Wang et al., "A Trusted Consensus Scheme for Collaborative Learning in the Edge AI Computing Domain," in IEEE Network, vol. 35, no. 1, pp. 204–210, January/February 2021, doi: https://doi.org/10.1109/MNET.011.2000249.
4. Guha, D. Samanta, A. Banerjee and D. Agarwal, "A Deep Learning Model for Information Loss Prevention From Multi-Page Digital Documents," in IEEE Access, vol. 9, pp. 80451–80465, 2021, doi: https://doi.org/10.1109/ACCESS.2021.3084841.
5. Rohan Kumar, Rajat Kumar, Pinki Kumar, Vishal Kumar, Sanjay Chakraborty, Prediction of Protein-Protein interaction as Carcinogenic using Deep Learning Techniques, 2nd International Conference on Intelligent Computing, Information and Control Systems (ICICCS), Springer, pp. 461–475, 2021.
6. Guha, A., Samanta, D. Hybrid Approach to Document Anomaly Detection: An Application to Facilitate RPA in Title Insurance. Int. J. Autom. Comput. 18, 55–72 (2021). https://doi.org/10.1007/s11633-020-1247-y
7. Lopamudra Dey, Sanjay Chakraborty, Anirban Mukhopadhyay. Machine Learning Techniques for Sequence-based Prediction of Viral-Host Interactions between SARS-CoV-2 and Human Proteins. Biomedical Journal, Elsevier, 2020.
8. Khamparia, A, Singh, PK, Rani, P, Samanta, D, Khanna, A, Bhushan, B. An internet of health things-driven deep learning framework for detection and classification of skin cancer using transfer learning. Trans Emerging Tel Tech. 2020;e3963. https://doi.org/10.1002/ett.3963
9. Jiawei Han and Micheline Kamber, Data Mining concepts and techniques , Morgan Kaufmann (publisher) from chapter-7 'cluster analysis', ISBN:978-1-55860-901-3, 2006.
10. Dunham, M.H., Data Mining: Introductory And Advanced Topics, New Jersey: Prentice Hall, ISBN-13: 9780130888921. 2003.
11. H.Witten, Data mining: practical machine learning tools and techniques with Java implementations San-Francisco, California : Morgan Kaufmann,ISBN: 978-0-12-374856-0 2000.
12. Kantardzic, M. Data Mining: concepts, models, method, and algorithms, New Jersey: IEEE press, ISBN: 978-0-471-22852-3, 2003.
13. Michael K. Ng, Mark Junjie Li, Joshua Zhexue Huang, and Zengyou He, " On the Impact of Dissimilarity Measure in k-Modes Clustering Algorithm ", IEEE transaction on pattern analysis and machine intelligence, vol.29, No. 3, March 2007.
14. NareshkumarNagwani and Ashok Bhansali, "An Object Oriented Email Clustering Model Using Weighted Similarities between Emails Attributes", International Journal of Research and Reviews in Computer science (IJRRCS), Vol. 1, No. 2, June 2010.
15. Oyelade, O.J, Oladipupo, O. O, Obagbuwa, I. C, "Application of k-means Clustering algorithm for prediction of Students' Academic Performance",(IJCSIS) International Journal of Computer Science and Information security,Vol.7,No. 1, 2010.

16. S.Jiang, X.Song, "A clustering based method for unsupervised intrusion detections" . Pattern Recognition Letters, PP.802–810, 2006.
17. Steven Young, ItemerArel, Thomas P. Karnowski,Derek Rose, University of Tennessee, "A Fast and Stable incremental clustering Algorithm", TN 37996, 7th International 2010.
18. Taoying Li and Yan Chen, "Fuzzy K-means Incremental Clustering Based on K-Center and Vector Quantization", Journal of computers, vol. 5, No.11, November 2010.
19. Tapas Kanungo , David M. Mount , "An Efficient k-Means Clustering Algorithm: Analysis and implementation IEEE transaction vol. 24 No. 7, July 2002.
20. Zuriana Abu Bakar, Mustafa Mat Deris and ArifahCheAlhadi, "Performance analysis of partitional and incremental clustering", SNATI, ISBN-979-756-061-6, 2005.
21. Xiaoke Su, Yang Lan, Renxia Wan, and Yuming, "A Fast Incremental Clustering Algorithm ", international Symposium on Information Processing (ISIP'09), Huangshan, P.R.China, August-21–23, pp:175–178,2009.
22. Kehar Singh, Dimple Malik and Naveen Sharma, "Evolving limitations in K-means algorithm in data Mining and their removal", IJCEM International Journal of Computational Engineering & Management, Vol. 12, April 2011.
23. Anil Kumar Tiwari, Lokesh Kumar Sharma, G. Rama Krishna, " Entropy Weighting Genetic k-Means Algorithm for Subspace Clustering ",International Journal of Computer Applications (0975–8887),Volume 7– No.7, October 2010.
24. K. Mumtaz, Dr. K. Duraiswamy, "An Analysis on Density Based Clustering of Multi Dimensional Spatial Data", Indian Journal of Computer Science and Engineering, Vol. 1 No 1, pp-8–12, ISSN : 0976-5166.
25. A.M.Sowjanya, M.Shashi, "Cluster Feature-Based Incremental Clustering Approach (CFICA) For Numerical Data, IJCSNS International Journal of Computer Science and Network Security, VOL.10 No.9, September 2010.
26. Martin Ester, Hans-Peter Kriegel, Jorg Sander, MichaelWimmer, Xiaowei Xu, "Incremental clustering for mining in a data ware housing", 24th VLDB Conference New York, USA, 1998.
27. SauravjyotiSarmah, Dhruba K. Bhattacharyya,"An Effective Technique for Clustering Incremental Gene Expression data" , IJCSI International Journal of Computer Science Issues, Vol. 7, Issue 3, No 3, May 2010.
28. Debashis Das Chakladar and Sanjay Chakraborty, Multi-target way of cursor movement in brain computer interface using unsupervised learning, Biologically Inspired Cognitive Architectures (Cognitive Systems Research), Elsevier, 2018.
29. Althar, R.R., Samanta, D. The realist approach for evaluation of computational intelligence in software engineering. Innovations Syst Softw Eng 17, 17–27 (2021). https://doi.org/10.1007/s11334-020-00383-2.
30. B. Naik, M. S. Obaidat, J. Nayak, D. Pelusi, P. Vijayakumar and S. H. Islam, "Intelligent Secure Ecosystem Based on Metaheuristic and Functional Link Neural Network for Edge of Things," in IEEE Transactions on Industrial Informatics, vol. 16, no. 3, pp. 1947–1956, March 2020, doi: https://doi.org/10.1109/TII.2019.2920831.
31. Debashis Das Chakladar and Sanjay Chakraborty, EEG Based Emotion Classification using Correlation Based Subset Selection, Biologically Inspired Cognitive Architectures (Cognitive Systems Research), Elsevier, 2018.
32. D. Samanta et al., "Cipher Block Chaining Support Vector Machine for Secured Decentralized Cloud Enabled Intelligent IoT Architecture," in IEEE Access, vol. 9, pp. 98013–98025, 2021, doi: https://doi.org/10.1109/ACCESS.2021.3095297.
33. CHEN Ning , CHEN An, ZHOU Long-xiang, "An Incremental Grid Density-Based Clustering Algorithm", Journal of Software, Vol.13, No.1,2002.
34. P. T. Gamage, M. Khurshidul Azad, A. Taebi, R. H. Sandler and H. A. Mansy, "Clustering Seismocardiographic Events using Unsupervised Machine Learning," 2018 IEEE Signal Processing in Medicine and Biology Symposium (SPMB), 2018, pp. 1–5, doi: https://doi.org/10.1109/SPMB.2018.8615615.

35. M. Elbattah, R. Carette, G. Dequen, J. -L. Guérin and F. Cilia, "Learning Clusters in Autism Spectrum Disorder: Image-Based Clustering of Eye-Tracking Scanpaths with Deep Autoencoder," 2019 41st Annual International Conference of the IEEE Engineering in Medicine and Biology Society (EMBC), 2019, pp. 1417–1420, doi: https://doi.org/10.1109/EMBC.2019.8856904.

36. K. P. Sinaga and M. Yang, "Unsupervised K-Means Clustering Algorithm," in IEEE Access, vol. 8, pp. 80716–80727, 2020, doi: https://doi.org/10.1109/ACCESS.2020.2988796.

37. K. Virupakshappa and E. Oruklu, "Unsupervised Machine Learning for Ultrasonic Flaw Detection using Gaussian Mixture Modeling, K-Means Clustering and Mean Shift Clustering," 2019 IEEE International Ultrasonics Symposium (IUS), 2019, pp. 647–649, doi: https://doi.org/10.1109/ULTSYM.2019.8926078.

38. G. Pu, L. Wang, J. Shen and F. Dong, "A hybrid unsupervised clustering-based anomaly detection method," in Tsinghua Science and Technology, vol. 26, no. 2, pp. 146–153, April 2021, doi: https://doi.org/10.26599/TST.2019.9010051.

39. L. R. Jiménez, "Web Page Classification based on Unsupervised Learning using MIME type Analysis," 2021 International Conference on COMmunication Systems & NETworkS (COMSNETS), 2021, pp. 375–377, doi: https://doi.org/10.1109/COMSNETS51098.2021.9352869.

40. A. A. Aktaş, A. T. Bayrak, O. Susuz and O. Tunalı, "An Application of Unsupervised Clustering Approaches in Customer Segmentation," 2020 4th International Symposium on Multidisciplinary Studies and Innovative Technologies (ISMSIT), 2020, pp. 1–6, doi: https://doi.org/10.1109/ISMSIT50672.2020.9254369.

41. M. A. Kabir and X. Luo, "Unsupervised Learning for Network Flow Based Anomaly Detection in the Era of Deep Learning," 2020 IEEE Sixth International Conference on Big Data Computing Service and Applications (BigDataService), 2020, pp. 165–168, doi: https://doi.org/10.1109/BigDataService49289.2020.00032.

42. S. S. Shaji and A. Varghese, "Unsupervised Segmentation of Images using CNN," 2020 International Conference on Smart Electronics and Communication (ICOSEC), 2020, pp. 403–406, doi: https://doi.org/10.1109/ICOSEC49089.2020.9215311.

43. A. K. Rai and R. K. Dwivedi, "Fraud Detection in Credit Card Data using Unsupervised Machine Learning Based Scheme," 2020 International Conference on Electronics and Sustainable Communication Systems (ICESC), 2020, pp. 421–426, doi: https://doi.org/10.1109/ICESC48915.2020.9155615.

Chapter 5
Research Intention Towards Incremental Clustering

5.1 Introduction

Incremental clustering is one of the most widely used incremental data mining techniques, where real-world datasets are continuously updated. This chapter mainly discusses the incremental problem and its solution approach. A brief explanation of the typical K-means and DBSCAN clustering algorithms has been discussed in Chap. 4. This chapter mainly derives the incremental nature of those popular K-means and DBSCAN clustering techniques as the solution of the problem.

5.2 Problem Definition

K-means and DBSCAN clustering are well-known, widely used and easy-to-handle techniques. They are very effective for static databases, but they are not suitable for dynamic databases where the data are frequently updated. In today's world, most of the datasets are dynamic in nature (such as stock market database, weather forecasting database, etc.). Because of rapidly changing data in dynamic databases, the existing K-means and DBSCAN clustering algorithms take a lot of time to resca the whole algorithm for each and individual new coming data in the database. However, time is a big matter for today's busy life [1, 2]. We must have to reduce the time as well as the effort. They not only take too much time, but it also has been realized that applying the existing algorithms frequently for updated databases may be too costly. Therefore, the existing clustering algorithms are not suitable for the dynamic environment [3].

© The Author(s), under exclusive license to Springer Nature Switzerland AG 2022
S. Chakraborty et al., *Data Classification and Incremental Clustering in Data Mining and Machine Learning*, EAI/Springer Innovations in Communication and Computing, https://doi.org/10.1007/978-3-030-93088-2_5

5.3   Solution Approach

To overcome the above problem, a new approach is to develop called incremental clustering algorithms that can measure the new cluster centres by directly computing the distance of the new data from the means of the existing clusters instead of rerunning the whole clustering algorithm. This concept is similar for both K-means and DBSCAN algorithms. Thus, it describes at what percent of delta change in the original database up to which incremental K-means or DBSCAN clustering behaves better than their existing algorithms [4].

These algorithms are mainly suitable for the dynamic environment. This new approach not only reduces time but also reduces effort and cost. Besides that, the result is bad for high-dimensional data with existing K-means clustering, but this incremental K-means algorithm performs better results with high-dimensional or multidimensional large data [5]. Figure 5.1 shows the methodology of incremental K-means algorithms.

This chapter explains that the actual K-means approach clustered the new coming data objects by repetitively running the K-means algorithm after assuming clusters initially. However, from "Fig. 5.1", it clears that the incremental K-means clustered the new data with the help of the means of existing clusters. Finally, it compares the processing time of these two approaches and concludes that the incremental approach takes less time than the actual one ($T_2 < T_1$) but up to some certain point of change in the database. The same approach must be followed in the DBSCAN clustering algorithm as well [6].

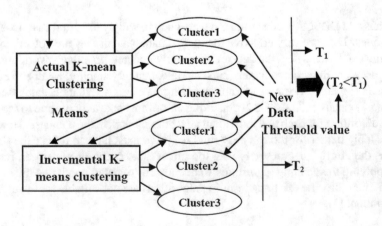

Fig. 5.1 Methodology of incremental K-means algorithms

5.4   Incremental Clustering

Incremental clustering mainly clusters the randomly new data into its similar group of clusters. The primary objective of this chapter is to cluster those newly inserted data in such a way that they take less amount of time to cluster than they take in the case of the actual one. In addition, time is always proportional to the effort, i.e., if the required time is decreased, the effort is also reduced proportionately, and another parameter, called cost, is also reduced [7, 8]. We describe below how the incremental K-means and DBSCAN are performed simply.

5.5   Incremental K-Means Clustering

For the implementation of a new incremental K-means clustering technique, a new incremental clustering algorithm is proposed, which is considerably superior in performance to the actual or existing K-means clustering approach. This method addresses the issues that occur from partitioning and clustering [9]. The probability of finding the global optimum may be improved by incremental clustering.

5.5.1   *Proposed Incremental K-Means Clustering Algorithm*

Explanation of the pseudocode: The following steps describe the above algorithm in details:

A. First, form a cluster based on the initial set of data items. Then run the actual K-means clustering algorithm on it, and compute the time it takes to form such clusters. Suppose it takes time T_1.
B. When the new data item is received, it can be assigned to one of the existing clusters, or a new cluster, or treat the new data as the outlier. The assignment is based on the distance between the new item and the existing cluster's centroids [10, 11]. In this chapter, we discuss when a new item is received, either it runs the whole K-means algorithm to cluster the new data, or it directly clustered the new item based on the closest mean between the new item and the means of the existing clusters. The incremental clustering follows the second approach to form clusters with newly received data. Suppose it takes time T_2.
C. Repeat step B until all the data samples is clustered.
D. Now, we compare the time T_1 and T_2 to find which approach is more effective, requires less effort and is less time-consuming.
E. Finally, determine the actual threshold value below which the incremental K-means clustering algorithm outperforms the actual K-means algorithm ($T_2 T_1$) (based on time and effort analysis).

Because the data in clusters is fixed, incremental clustering differs from partitioned clustering.

5.5.2   Proposed Model of Incremental K-Means Clustering

New transactions (records/rows) are added to dynamic databases as time passes. It's an incremental algorithm that's been utilized to solve this problem. The suggested algorithm calculates the percentage of the original database's size that can be added to the new database. Now, there could be two scenarios:

1. If the original database has changed by more than x percent, it is preferable to use the previous result.
2. Rerun the algorithm if the percentage is greater than x percent [12].

Solution: As a result, incremented K-means will be faster to execute than earlier K-means algorithms because the number of database scans will be reduced.

"Fig. 5.2" describes the main aim of this chapter clearly. First, the existing K-means clustering algorithm (developed in Java) is applied to the original air

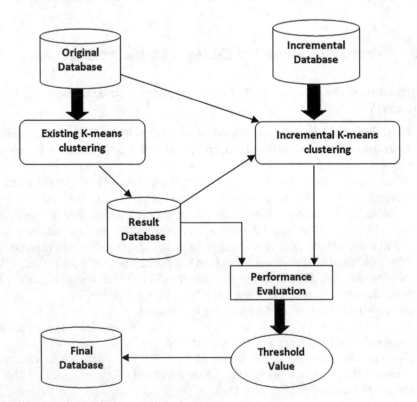

Fig. 5.2 Proposed model of incremental K-means clustering

pollution database, and the result is stored in a result database using MySQL. If new data is received and inserted into that existing database, this is known as an incremental database. After gathering the essential information from the result database, the incremental K-means clustering method is used for that incremental data. As a result, the new data is directly integrated into the old database, eliminating the need to run the K-means algorithm repeatedly. Finally, the two outcomes are compared, and the performance and the correct threshold value is evaluated. The mathematical examples in the next section helps to clarify this concept [13, 14].

5.5.3   Illustrative Examples of Incremental K-Means Clustering

The proposed model can be explained using the two examples below:

Example 1. Assume that you have a collection of data objects, such as A(15), B(7), C(8), D(11), E(5), F(14), G(3), and H(1). Assume that points A, E, and H are three cluster centers. Form clusters properly using the K-means algorithm. Suppose two new data I(17) and S(9) are inserted, and also two data items, D(11), and F(14) are deleted later. Then show how this algorithm will behave?

Sol: Suppose the initial cluster centres are A(15), E(5), and H(1). Computation is done below using the Manhattan distance metric $(D = |x_1-x_2| + |x_2-x_3| + ...)$. Table 5.1 shows values of data items after first iteration.

First Iteration:

Second Iteration:

Now,   #items   Mean
Cluster 1 = {A(15),D(11),F(14)}   =3   = 13.3
Cluster 2 = {B(7),C(8),E(5)}   =3   = 6.7
Cluster 3 = {G(3),H(1)}   =2   = 2

Table 5.1 Values of data items after first iteration

Data items	A(15)	E(5)	H(1)	New clusters
A(15)	0(min)	10	14	1
B(7)	8	2	6	2
C(8)	7	3	7	2
D(11)	4	6	10	1
E(5)	10	0	4	2
F(14)	1	9	13	1
G(3)	12	2	2	3
H(1)	14	4	0	3

Based on these three means, the items' distances and their group of clusters can be computed [15]. Table 5.2 projects values of new groups of cluster items. If there is no change in the 2nd iteration of the data, then this algorithm is terminated. This algorithm is also known as 3-means clustering. According to the mean value, the new groups of cluster items are,

No Change

Insertion:
Two new data items, I(17), and S(9), are inserted, and then in the first approach, those data are clustered directly after comparing with the means of existing clusters using the Manhattan distance metric. Such as

I(17) = > 3.7(min)	10.3	15	= > directly entered into the cluster 1
S(9) = > 4.3	2.3	7	= > directly entered into the cluster 2

However, in the second approach, if the K-means algorithm is run again with those two new data items, and it results as follows shown in Table 5.3.

The result is the same, but the second approach is more time-consuming and requires more effort compared to the first approach.

Deletion:
Now, if two items, D(11), and F(14), are deleted from the existing database, then in the first approach, after calculating from clusters new means(after deletion), expressed in Table 5.4,

Now	#items
Cluster 1 = {A(15)}	= 1
Cluster 2 = {B(7),C(8),E(5)}	= 3
Cluster 3 = {G(3),H(1)}	= 2

Table 5.2 Values of new groups of cluster items

Data items	New clusters
A(15)	1
B(7)	2
C(8)	2
D(11)	1
E(5)	2
F(14)	1
G(3)	3
H(1)	3

Table 5.3 Values of data item after the insertion of I(17) and S(9)

Data items	A(15)	E(5)	H(1)	New clusters
A(15)	0(min)	10	14	1
B(7)	8	2	6	2
C(8)	7	3	7	2
D(11)	4	6	10	1
E(5)	10	0	4	2
F(14)	1	9	13	1
G(3)	12	2	2	3
H(1)	14	4	0	3
I(17)	2	13	16	1
S(9)	6	4	8	2

Table 5.4 Values of data item after the deletion of D(11) and F(14)

Data items	A(15)	E(5)	H(1)	New clusters
A(15)	0(min)	10	14	1
B(7)	8	2	6	2
C(8)	7	3	7	2
E(5)	10	0	4	2
G(3)	12	2	2	3
H(1)	14	4	0	3

Now,

	#items	Mean
Cluster 1 = {A(15)}	= 1	= 15
Cluster 2 = {B(7),C(8),E(5)}	= 3	= 6.7
Cluster 3 = {G(3),H(1)}	= 2	= 2

No Change in the Cluster

Here also, the 2nd approach is time-consuming and requires more effort compared to the first one.

Example 2. Sometimes the mean of a cluster depends on the dimensions of its using database [16]. Suppose a multidimensional database has four attributes, so each cluster of that database must produce four centroids or means. Suppose an air pollution database has four attributes SPM, RPM, nitrogen dioxide (NO_2), and sulfur dioxide (SO_2). Assume the initial number of cluster is 3. Table 5.5 shows the means of each cluster-based and also assumes that after the first iteration, the means of each cluster are:

Now, if a new data is entered into the existing database with value SPM = 21, RPM = 8, NO_2 = 9, and SO_2 = 12, then it first compares each of its attribute's distance with the attributes of the existing cluster with the help of distance metric (Euclidean metric). And it will enter into that cluster where the distances are minimum. Such as

Table 5.5 Means of each cluster based

	Cluster 0	Cluster 1	Cluster 2
SPM	24	32	15
RPM	22	42	20
NO₂	12	32	9
SO₂	14	27	12

$$\text{Cluster } 0 = \sqrt{(24-21)+(22-8)+(12-9)+(14-12)} = 4.7$$

$$\text{Cluster } 1 = \sqrt{(32-21)+(42-8)+(32-9)+(27-12)} = 9.11$$

$$\text{Cluster } 2 = \sqrt{(21-15)+(20-8)+(9-9)+(12-12)} = 4.2 \text{ (minimum)}$$

Therefore, without rerunning the algorithm, the new data item should be instantly inserted into "Cluster 2". As a result, it saves us both time and effort.

5.5.4   Benefits and Applications of Incremental K-Means Clustering

The actual K-means technique is unsuitable for a big, often updated multidimensional database. In that situation, gradual clustering is a much better option. The incremental technique saves a lot of time, effort, and money, whereas the old system already has several flaws, which are most noticeable in a large dynamic databases. It deals with the problem of scanning the entire database again when certain records are added to existing data. Due to the need to rescan the entire database, the time complexity is extremely high. The current system is inefficient in terms of time and effort. As a result, the new incremental clustering algorithm is more suited for application in a large multidimensional dynamic database [17, 18].

Incremental K-means clustering algorithm is widely used in educational research, market research, medicine and biomedical, neuroscience, software evolution, social network analysis, image segmentation, data mining, climatology, the grouping of shopping items, fraud detection and, etc. Today, a system is given to assess student results using cluster analysis and basic statistical techniques to organize their scores according to their performance level [19]. In this example, the K-means clustering technique is often utilized, and we can complete all of those tasks significantly faster and with much better results by using incremental K-means.

5.6   Incremental DBSCAN Clustering

A new density-based incremental clustering algorithm (DBSCAN) is introduced to remove the drawbacks of incremental K-means clustering. It is one of the most common clustering algorithms and also most cited in the scientific literature. It follows the almost same method to cluster the incremental data in a dynamic environment like K-means, but it can handle noisy data. The performance comparison is performed between incremental K-means and incremental DBSCAN algorithms in the "Experiment and Results" section.

5.6.1   Explanation of Pseudocode

DBSCAN is sometimes used on a dynamic database that is frequently updated by data insertion or deletion. The clustering detected by DBSCAN must be updated after insertions and deletions to the database. The probability of finding the global optimum may be improved by incremental clustering. We initially build clusters based on the initial objects and a given radius (eps) and a minimum number of points in this method (Minpts) [20]. As a result, we receive several clusters that meet the criteria, and some outliers. When new data is inserted into the existing database, we must now use DBSCAN to update our clusters. We first compute the means between each cluster's core object and the new data, and then place the new data into the cluster with the shortest mean distance. New data that isn't incorporated into any clusters is considered noise or an anomaly. Outliers that meet the Minpts and eps criterion can sometimes combine to form new clusters [21].

5.6.2   Proposed Model of Incremental DBSCAN Clustering

The proposed model may be presented using "Fig. 5.3", which shows how the DBSCAN algorithm is applied to the original database, and the results are stored in a MySQL or other database. The incremental DBSCAN algorithm is applied to the incremental dataset, and the results are compared, and the performance is evaluated [22].

5.6.3   Benefits of Incremental DBSCAN Clustering

For a big multidimensional database that is constantly updated, the real DBSCAN technique is insufficient. In that situation, gradual clustering is a much better option. Although the previous system already has some shortcomings, the

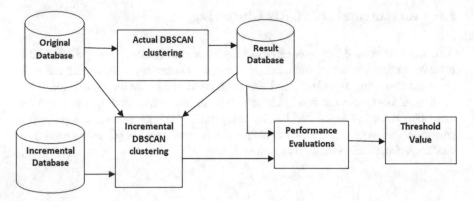

Fig. 5.3 Proposed model of incremental DBSCAN clustering

incremental system saves a lot of time and work, primarily encountered in large dynamic datasets by the existing system [23, 24]. It deals with the problem of scanning the entire database again when certain records are added to existing data. Due to the need to rescan the entire database, the time complexity is extremely high. The current system is inefficient in terms of time and effort. As a result, the new incremental clustering algorithm is more suited for application in a large multidimensional dynamic database [25].

After the performance evaluation of the incremental K-means and DBSCAN clustering algorithms properly, a performance comparison is made. By the help of this comparison, it can be concluded which incremental algorithm is better on the incremental database. Here air pollution database is used as an incremental database in the experiment [26].

5.7 Effects of Cluster Metadata on Incremental Clustering

Every cluster analysis technique depends on some necessary features or characteristics of clustering. Those features can change based on the conditions. These features or characteristics of clustering are known as "cluster metadata". This incremental K-means and DBSCAN clustering methods are also based on the cluster metadata [27]. The behavior of incremental clustering is also changed due to the effects of metadata. Clustering depends on five types of cluster metadata, such as

(a) *Cluster centre (C)*
(b) *Cluster radius (R)*
(c) *Number of cluster (K)*
(d) *Number of data items (N) in cluster*
(e) *Mean value of cluster (M)*

How this cluster metadata affects the actual and incremental clustering methods discussed below:

(a) *Cluster centre*

Cluster center plays a major role in building new clusters in a dynamic environment. When some new data arrive, the cluster centre measures the distance of the new data from itself using ideal distance measure function and receives the least distance data based on the minimum distance. Cluster center must be fixed [28, 29]. In Example 1, we have described how the cluster center clusters the newly received data using the K-means clustering algorithm.

(b) *Cluster radius*

Cluster radius is also important metadata that plays a useful role, especially in DBSCAN clustering. Cluster radius is mainly used to detect noisy data or outliers. It makes the clustering technique more efficient. The data items beyond the limitation of cluster radius are called outliers. Thus the distance of the new coming data are measured and found whether they are beyond the cluster radius or not [30]. It is challenging to calculate the accurate cluster radius in the case of K-means clustering, but in the case of DBSCAN clustering, the cluster radius must be predefined. The role of cluster radius is elaborately discussed in the illustrative examples section of DBSCAN clustering.

(c) *Number of clusters*

In the case of K-means clustering, K defines the total number of clusters. This K must be predefined. If the number of cluster increases, the more number of data items will be included in those clusters, and as a result, the number of outliers will be decreased [31].

(d) *Number of data items (N) in cluster*

In the case of K-means clustering, the total number of clusters is fixed. If the numbers of data items increase, then insert those data into existing predefined clusters, or treat those data as the outliers. But in the case of DBSCAN clustering, if numbers of data items increase, it simply inserts those items based on its cluster radius value, and sometimes several newly inserted data can form new clusters.

(e) *Mean value of cluster*

.

Overlapping Issues

Sometimes a new concept appears when two clusters are merged or overlapped by some portion, and they have only one common center in that common portion, and then those clusters are treated as a single cluster [32]. Due to this overlapping, they cover much more numbers of objects in the data space. This technique is also referred to as "multilevel clustering". The following "Fig. 5.4" discusses this concept.

This chapter describes the introduction of the two new proposed incremental clustering algorithms. This section also explains the efficiency of those two incremental clustering algorithms with the help of mathematical explanation. Based on the proposed algorithms and proposed concepts, implementations of these two clustering algorithms are clearly explained in the next section [33]. The next section also provides the necessary simulation to analyze the performance of incremental

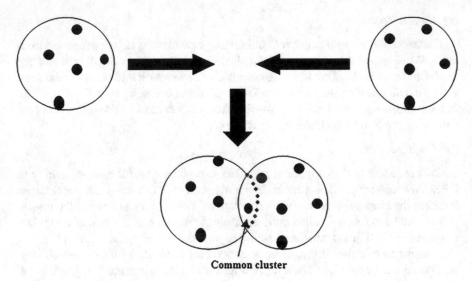

Fig. 5.4 Multilevel clustering

K-means and DBSCAN clustering algorithms, and finally, it describes performance comparisons in between incremental K-means and DBSCAN clustering algorithms.

5.8   Experiment and Result Analysis

The necessary tools to achieve the experiment results are as follows.

5.8.1   Air Pollution Database

This analysis is based on the observation of the air pollution database which has been collected from "West Bengal Air Pollution Control Board" and the URL is-"http://www.wbpcb.gov.in/html/airqualitynxt.php" [18]. The detail database format is shown in "Table 5.6".

5.8.2   Java Platform

"Java" stands for "just avail vital abstraction". Java is a platform-independent, object-oriented, simple, portable, and highly reliable powerful programming language developed by Sun Microsystems in 1991. It provides a way of modularising programs by creating partitioned memory area for both data and functions [34]. Java has several noticeable features, such as encapsulation, data abstraction, inheritance,

Table 5.6 Original air pollution database

Date	SPM	RPM	SO$_2$	NO$_X$
1/1/2009	357	183	12	95
2/1/2009	511	289	14	125
3/1/2009	398	221	10	101
4/1/2009	358	191	11	97
5/1/2009	329	175	11	101
....	....	....	......	....
1/2/2009	376	197	10	105
2/2/2009	314	151	10	85
3/2/2009	285	154	8	96
4/2/2009	352	195	8	93
....	....	....	......	....

polymorphism, and dynamic binding. This chapter uses Java 1.5 compiler to simulate the proposed incremental clustering algorithms. A higher version of Java (Java 7.0) is not used because of generics problem.

5.8.3 Weka Framework

Weka is the other open-source API's developed by researchers at the University of Waikato in New Zealand. Weka is a Java-based machine learning tool. Weka has three modes of operation

(i) GUI
(ii) Command line
(iii) Java API

A collection of open-source algorithms:

(i) Preprocessing
(ii) Classifiers
(iii) Clustering
(iv) Association rule

In this chapter, Weka 3.6 is used. Weka can be easily available from the website – "http://www.cs.waikato.ac.nz/ml/weka/". Weka is mainly used for performing some data mining-related operations.

"MySQL" is mainly used to construct databases and also used to store the results of clustering. Some important functions of this research work are given below in Java.

(i) **Actual K-means algorithm implementation**

```
// Reading original file
publicactualkmeans(String filename)throws Exception
   {
String[] options;
    Instances data = new Instances(new BufferedReader(new FileReader(filename)));
System.out.println("\n--> K-means clutering");


.................................................................................................
// Running simple K-means from Weka

SimpleKMeanskMeans = new SimpleKMeans();
kMeans.setNumClusters(5);
System.out.println(ClusterEvaluation.evaluateClusterer(kMeans,options));
// calculate number of iteration
kMeans.buildClusterer(data);
   String kmeansresult=kMeans.toString();
   String []strArr=kmeansresult.split("\n");
    String supportLine=strArr[4];
   String []suppV = supportLine.split(":");
        String sup = suppV[1];
        sup = sup.split("\\(")[0];
        support1 = Double.parseDouble(sup);
System.out.println(support1);


.............................................................................................
// splitting and storing means
String[] result = "321.376238,164.366337,10.128713,92.415842".split(",");
for (int x=0; x<result.length; x++)
System.out.println(result[x]);
System.out.println("\n");
.................................................................................................
// Required time calculation
public long timerequirement()throws Exception
        {
        long start;
        long end;
        start=System.currentTimeMillis();
        newactualkmeans("d:\\original air pollution data.arff");
        end=System.currentTimeMillis();
        longTimerequired=(end - start);
        System.out.println("---The needed time for the actual k-means on original air
pollution data.arff file: "+Timerequired+" ms");
        returnTimerequired; } }
```

5.8.4   Explanation of the Typical K-Means Algorithm Code

In the first step, the air pollution database (4D data) is stored in the .arff file, which can be read with the help of "BufferedReader" class and "FileReader" method, and is stored into data variable. Then the simple K-means algorithm runs on that data after assuming five clusters initially [35]. As a result, five means (each mean have four values) come out and store those resultant means into an array through splitting and parsing methods. At last, the total time is calculated for the typical K-means algorithm.

(ii) **Incremental K-means Algorithm Implementation**

```
// Reading incremental data
ArrayList<String> list = showLines("d:\\incremental air pollution data.arff", 7,507);
..............................................................................................................
// Covert incremental data into double format

doubledarray[][]=new double[list.size()][4];
        System.out.println("\n");
        System.out.println("New incoming data are converted into double");
        for (int r = 0; r<list.size(); r++)
          {
                for (int c = 0; c<4; c++)
                  {
                        darray[r][c]=Double.parseDouble(twoDArray[r][c]);
                        //System.out.print(" Result is:"+darray[r][c]);
                        }
                //System.out.print("\n");
          }

//minimum distance calculation directly from mean values

4. for(int m=0; m<darray.length;m++)
       {
for(int n=0;n<darray[m].length;n++)
            {
        double distance1 = CalculateDistance ( d0, d1, d2, d3,darray[m][n]);
        double distance2 = CalculateDistance ( e0, e1, e2, e3,darray[m][n]);
        double distance3 = CalculateDistance ( f0, f1, f2, f3,darray[m][n]);
        double distance4 = CalculateDistance ( g0, g1, g2, g3,darray[m][n]);
        double distance5 = CalculateDistance ( h0, h1, h2, h3,darray[m][n]);
doubleminDist = minimum(distance1,distance2,distance3,distance4,distance5);
if(minDist == distance1)
        {
finalResult[m] = 1;
        }
else if(minDist == distance2)
        {  finalResult[m] = 2;
        }
else if(minDist == distance3)
```

```
      {   finalResult[m] = 3;
      }
else if(minDist == distance4)
      {
finalResult[m] = 4;
      }
else if(minDist == distance5)
      {
      finalResult[m] = 5;
      }
      System.out.println("The Data is belonging to cluster : " + finalResult[m]);
      } }
```
....

// **Body of CalculateDistance function**

```
public static double CalculateDistance(double x1, double  y1,double  z1, double k1,double
pass)
{
doublediffx= x1 - pass;
doublediffy= y1 - pass;
doublediffz = z1 - pass;
doublediffk = k1 - pass;
doublediffx_sqr = (diffx*diffx);
doublediffy_sqr = (diffy*diffy);
doublediffz_sqr = (diffz*diffz);
doublediffk_sqr = (diffk*diffk);
doubleFdistance = Math.sqrt(diffx_sqr + diffy_sqr + diffz_sqr + diffk_sqr);
returnFdistance;
}
```
....

// **Calculate minimum Euclidean distance**

```
public static double minimum(double E1,double E2,double E3,double E4,double E5)
   {
      double min = 0;
         if (E1<E2 && E1<E3 && E1<E4 && E1<E5)
      {
            System.out.println("The minimum distance of the current data is:"+E1);
            min = E1;
      }
      else if(E2<E1 && E2<E3 && E2<E4 && E2<E5)
      {
```

```
                System.out.println("The minimum distance of the current data is:"+E2);
                min = E2;
          }
        else if(E3<E1 && E3<E2 && E3<E4 && E3<E5)
        {
                System.out.println("The minimum distance of the current data is:"+E3);
                min = E3;
          }
        else if(E4<E1 && E4<E2 && E4<E3 && E4<E5)
        {
                System.out.println("The minimum distance of the current data is:"+E4);
                min = E4;
          }
else if(E5<E1 && E5<E2 && E5<E3 && E5<E4)
        {
                System.out.println("The minimum distance of the current data is:"+E5);
                min = E5;
          }
        return min;
  }
..................................................................................................

// Finally, store the result into a final database using MySQL.
```

5.8.5   Explanation of the Incremental K-Means Algorithm Code

At first, the new coming data into the database are read and then convert those incremental newly received data into double format [36]. Then by the help of "CalculateDistance" function, the minimum distance of the new incremental data is directly calculated from the existing cluster means (calculated earlier) using the Euclidean distance metric method. At last the new data are assigned to the nearest existing clusters based on their smallest distances.

(iii) **Incremental DBSCAN Clustering Algorithm Implementation**

The implementation of incremental DBSCAN is almost same as the incremental K-means Clustering. But there is only one difference which is based on the handling of noisy data or outliers. When the outliers are detected by the incremental DBSCAN algorithm, it simply grouping those outliers together into one cluster after satisfying the conditions of eps and Minpts.

....

// Handling of outliers (only for one data)
Public static double Outlierhandling(double re, double E1)
{
if((re>eps) || (E1!=min))   **//re is radius of cluster**
    {   **// min calculated same as actual K-means**
out=E1;
    }
return E1;
}
....
…
Then E1 must be entered into a new cluster, which is called 'the cluster of outliers'.
Create New-cluster ──▶ C_N
&
C_N ◀──────── E1.

Explanation of the Incremental DBSCAN Algorithm Code

Capability of handling outlier data by the incremental DBSCAN clustering algorithm is shown in the above coding section. If the data E_1 is not minimum and the radius of the cluster is greater than eps, then E_1 will be treated as outlier, and E_1 will be entered into the newly created cluster C_N. If the number of outliers is more than the Minpts, then all those outliers will enter into that cluster C_N, and finally C_N will be called *cluster of outliers*.

5.8.6   Result Analysis

This section analyses and expresses how these two incremental clustering algorithms can be applied to change database frequently, and how these algorithms perform better than the existing clustering algorithms concerning the time-up some specific point of changes in the incremental database. In the performance evaluation, both techniques involve the computation of centroids, where these centroids will be used to cluster the data. In the actual K-means clustering, the algorithm is applied to the air pollution dataset and forms clusters based on the nearest distance of the data from predefined centroids. However, in a dynamic environment, incremental K-means clustering is applied when newly received data is incorporated into the database. This technique performs its operations to the existing clusters and makes clusters of the newly received data directly by using the nearest distance between the new data and the centroids of the existing clusters. Both the techniques use the Euclidean distance measure function in this experiment [37].

At first, initialise the total number of clusters to five, and then the actual K-means clustering algorithm is running on a four-dimension attributes-based air pollution database. Table 5.7 shows the means of five clusters, and Table 5.8 expresses different parameters of the actual K-means clustering algorithm. So, each cluster consists of four objects. The result are stored in two different databases, which are described below:

This research defines the means of the five clusters after applying existing K-means clustering algorithm on the air pollution database. Here the required time is measured using "currentTimeMillis()" method of Java. After measuring the time for the change of data in the database, "Table 5.9" and "Fig. 5.5" can be shown.

Table 5.7   Means of five clusters

clusterid	clustSPMmean	clustRPMmean	clustSOmean	clustNOmean
cluster0	321.376238	164.366337	10.128713	92.415842
cluster1	252.600000	118.562500	8.425000	72.187500
cluster2	93.458824	36.176471	5.158824	41.523529
cluster3	165.196721	75.983607	6.704918	57.04918
cluster4	388.943182	202.022727	12.034091	107.102273

Table 5.8   Different parameters of the actual K-means clustering algorithm

clusternumber	distancefunction	clusteriteration	squareError
5	Euclidean distance	35.0000	12.53647

Table 5.9   Time vs. data in actual K-means clustering

Original data	Time (ms)
1000	156 ms
1100	172 ms
1200	172 ms
1300	187 ms
1400	188 ms
1500	188 ms
1600	203 ms
....	....

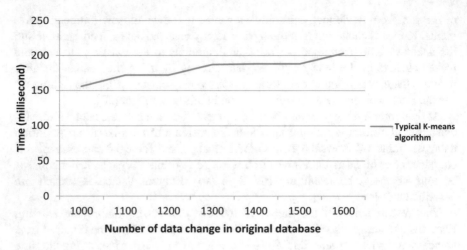

Fig. 5.5 Graph for actual K-means result

Table 5.10 Time vs. incremented data in incremental K-means clustering

Incremental data	Time (ms)
100	47
200	94
300	125
400	172
500	178
600	218
……..	……..

Figure 5.5 depicts how the time slows down as the amount of data in the original database grows. When new data is introduced into the old database, the proposed incremental K-means clustering technique is applied to the new data [38]. By comparing the new data to the means of existing clusters, this technique clustered the new data without re-executing the K-means algorithm. The relationship between the needed time and the new incremented data is shown in "Table 5.10" and "Fig. 5.6" below.

Figure 5.6 depicts how the time in the incremental database rapidly grows as the amount of data increases. After integrating the above two results, it is now simple to determine for what percentage (percent) of delta (δ) change in the database the incremental K-means clustering performs better than the true K-means clustering. To begin, use the formula below to determine all of the database's delta changes.

$$\%\delta \text{ change in DB} = \frac{(NEW\ DATA - OLD\ DATA)}{OLD\ DATA} \times 100$$

(5.2)

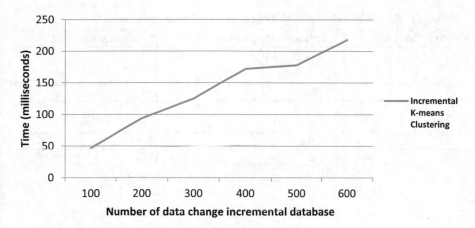

Fig. 5.6 Graph for incremental K-means result

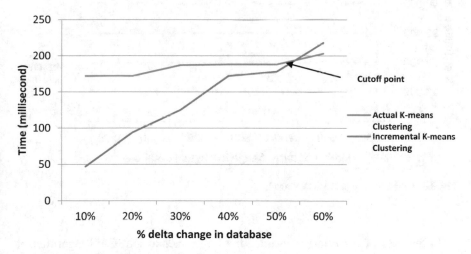

Fig. 5.7 Graph for actual K-means vs. incremental K-means

Suppose the database has 1000 data initially, and at each stage, 100 new data are inserted into the old database. Then the percent (%) of delta (δ) change in the database can be calculated in the following manner and expressed in Table 5.11.

From the above "Fig. 5.7", it can be easily mentioned that the threshold value up to which the proposed K-means clustering behaves better than the existing one is 57% [threshold value = 57%]. However, after that threshold value, the actual clustering technique behaves better than the incremental clustering [39].

Table 5.11 Time vs. % δ change in DB for both actual and incremental K-means

Actual Time (ms)	% δ change in the database	Incremental time (ms)
172	$\delta_1 = \dfrac{(1100 - 1000)}{1000} \times 100 = 10\%$	47
172	$\delta_2 = 20\%$	94
187	$\delta_3 = 30\%$	125
188	$\delta_4 = 40\%$	172
188	$\delta_5 = 50\%$	178
203	$\delta_6 = 60\%$	218
……..	……….	…….

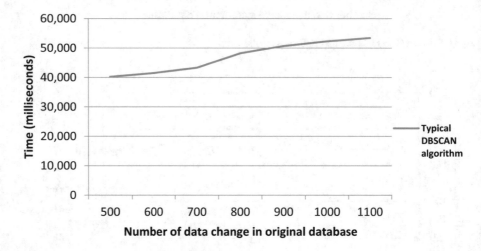

Fig. 5.8 Graph for actual DBSCAN result

In the case of performance evaluation of incremental DBSCAN algorithm, it depends not only on the calculation of means of the new data but also on its minimum number of points Minpts and its radius eps (discussed earlier). Figure 5.8 shows a graph for the actual DBSCAN result. The characteristic graphs below indicate the behavior change of existing DBSCAN and incremental DBSCAN algorithms (Fig. 5.8).

5.8.7   Clustering vs. Incremental Clustering

Clustering is a concept of making groups of similar data on static databases. Clustering cannot be applied on dynamic databases where data are frequently changed [40, 41]. Clustering is very simple to use, but this concept is not suitable

Table 5.12   Performance comparisons between clustering and incremental clustering

Characteristics	Clustering	Incremental clustering
Static Database(SDB)	Suitable	Suitable
Dynamic database(DDB)	Inefficient	Efficient
Time complexity(SDB)	Good	Average
Time complexity(DDB)	Average	Good (up to a certain threshold value in DB)
Memory size	Less required	More required
Cost	High	Low

nowadays because most of the databases (data warehouses, WWW, and so on) used today are dynamic, meaning data inserted into those databases and deleted from those databases are random in nature. For this purpose, clustering is inefficient. It is only efficient with static databases where data are static or fixed in nature. Clustering produces a large time complexity if it is applied to large dynamic databases [42, 43].

A new clustering concept has been developed to prevent this main drawback of clustering. It is referred to as "incremental clustering". This is the advanced concept of the existing clustering [44]. This chapter has already implemented and discussed the incremental ideas of K-means and DBSCAN clustering elaborately [45, 46]. This is the new revolution in the data mining field. With the help of this incremental clustering technique, one can easily handle a large dynamic database efficiently. An incremental clustering is more time-efficient, which means it consumes less time than the existing clustering. Thus, it requires less effort and less cost. "Table 5.12" describes the differences of characteristics between clustering and incremental clustering.

This chapter describes the application or performance of incremental K-means and incremental DBSCAN algorithms on a particular incremental database and compares their performances with their typical clustering algorithms [47]. It also explains the performance comparisons between incremental K-means and incremental DBSCAN clustering algorithms [48].

5.9   Conclusion

Using incremental concepts in the clustering method was the main aim of this chapter. Here mainly two clustering algorithms, K-means and DBSCAN, are selected for incremental implementation. In this chapter incremental concepts of these two clustering algorithms are implemented, and it also compares their incremental characteristics. Incremental clustering technique is used to overcome some of the problems related to the existing K-means and DBSCAN clustering in a dynamic environment. The actual K-means and DBSCAN are the two most popular clustering techniques, but they are not suitable in a dynamic environment where changes frequently happen in the database. The typical clustering algorithms take a lot of time, working with dynamic databases. Therefore, the main goal of this chapter is to develop and implement two new improved incremental K-means and DBSCAN clustering algorithms that can perform suitable clustering of incremental data in dynamic environment up to some certain point of change in the original database. It will make those algorithms more time-efficient and cost-efficient up to some certain threshold value in the database. This chapter evaluates the performance of both the K-means and DBSCAN clustering algorithms through basic programming tools and concepts. At last, their performances are also comparedto prove which algorithm provides better results on an incremental database.

5.10 Exercise

Q1. What are the various metrics involved to show the benefits of incremental clustering over the existing clustering techniques? How are they showed the advantages?
Q2. What are the disadvantages of K-means and DBSCAN clustering algorithms on handling static databases?
Q3. What do you understand by percent (%) of delta change on a database? Explain.
Q4. What are the various kinds of data on which incremental clustering can be applied suitably?

References

1. Eshref Januzaj, Hans-Peter Kriegel, Martin Pfeifle, "Towards Effective and Efficient Distributed Clustering", Workshop on Clustering Large Data Sets (ICDM2003), Melbourne, FL, 2003.
2. S. Jiang, X. Song, "A clustering based method for unsupervised intrusion detections" . Pattern Recognition Letters, PP. 802–810, 2006.
3. Guha A., D. Samanta, A. Banerjee and D. Agarwal, "A Deep Learning Model for Information Loss Prevention From Multi-Page Digital Documents," in IEEE Access, vol. 9, pp. 80451–80465, 2021, doi:https://doi.org/10.1109/ACCESS.2021.3084841.
4. A.M. Sowjanya, M. Shashi, "Cluster Feature-Based Incremental Clustering Approach (CFICA) For Numerical Data," IJCSNS International Journal of Computer Science and Network Security, VOL. 10 No. 9, September 2010.
5. Air-pollution database, WBPCB, URL: 'http://www.wbpcb.gov.in/html/airqualitynxt.php'.
6. Althar, R.R., Samanta, D. The realist approach for evaluation of computational intelligence in software engineering. Innovations Syst Softw Eng 17, 17–27 (2021). doi:https://doi.org/10.1007/s11334-020-00383-2.
7. Anil Kumar Tiwari, Lokesh Kumar Sharma, G. Rama Krishna, "Entropy Weighting Genetic k-Means Algorithm for Subspace Clustering", International Journal of Computer Applications (0975– 8887), Volume 7– No. 7, October 2010.
8. Aristidis Likasa, Nikos Vlassis, Jakob J. Verbeek,"The global k-means clustering algorithm", the journal of the pattern recognition society, Pattern Recognition 36 (2003) 451–461, 2002.

9. B. Naik, M. S. Obaidat, J. Nayak, D. Pelusi, P. Vijayakumar and S. H. Islam, "Intelligent Secure Ecosystem Based on Metaheuristic and Functional Link Neural Network for Edge of Things," in IEEE Transactions on Industrial Informatics, vol. 16, no. 3, pp. 1947–1956, March 2020, doi:https://doi.org/10.1109/TII.2019.2920831.

10. Carlos Ordonez and Edward Omiecinski, "Efficient Disk-Based K-Means Clustering for Relational Databases", IEEE transaction on knowledge and Data Engineering, Vol. 16, No. 8,August 2004.

11. Carlos Ordonez, "Clustering Binary Data Streams with K-means", San Diego, CA, USA. Copyright 2003, ACM 1- 58113-763-x, DMKD'03, June 13, 2003.

12. CHEN Ning, CHEN An, ZHOU Long-xiang, "An Incremental Grid Density-Based Clustering Algorithm", Journal of Software, Vol. 13, No. 1, 2002.

13. D. Samanta et al., "Cipher Block Chaining Support Vector Machine for Secured Decentralized Cloud Enabled Intelligent IoT Architecture," in IEEE Access, vol. 9, pp. 98013–98025, 2021, doi:https://doi.org/10.1109/ACCESS.2021.3095297.

14. Data Mining concepts and techniques by Jiawei Han and Micheline Kamber, Morgan Kaufmann (publisher) from chapter-7 'cluster analysis', ISBN:978-1-55860-901-3, 2006.

15. Debashis Das Chakladar and Sanjay Chakraborty, EEG Based Emotion Classification using Correlation Based Subset Selection, Biologically Inspired Cognitive Architectures (Cognitive Systems Research), Elsevier, 2018.

16. Dunham, M.H., Data Mining: Introductory And Advanced Topics, New Jersey: Prentice Hall, ISBN-13: 9780130888921. 2003.

17. Govender, P., & Sivakumar, V. (2020). Application of k-means and hierarchical clustering techniques for analysis of air pollution: A review (1980–2019). Atmospheric pollution research, 11(1), 40–56.

18. Guha, A., Samanta, D. Hybrid Approach to Document Anomaly Detection: An Application to Facilitate RPA in Title Insurance. Int. J. Autom. Comput. 18, 55–72 (2021). doi:https://doi.org/10.1007/s11633-020-1247-y

19. H. Witten, Data mining: practical machine learning tools and techniques with Java implementations San-Francisco, California: Morgan Kaufmann, ISBN: 978-0-12-374856-0 2000.

20. Jahwar, A. F., & Abdulazeez, A. M. (2020). Meta-heuristic algorithms for k-means clustering: A review. PalArch's Journal of Archaeology of Egypt/Egyptology, 17(7), 12002–12020.

21. K. Mumtaz, Dr. K. Duraiswamy, "An Analysis on Density Based Clustering of Multi Dimensional Spatial Data", Indian Journal of Computer Science and Engineering, Vol. 1 No 1, pp-8–12, ISSN: 0976-5166.

22. K. Wang et al., "A Trusted Consensus Scheme for Collaborative Learning in the Edge AI Computing Domain," in IEEE Network, vol. 35, no. 1, pp. 204–210, January/February 2021, doi:https://doi.org/10.1109/MNET.011.2000249.

23. Kantardzic, M. Data Mining: concepts, models, method, and algorithms, New Jersey: IEEE press, ISBN: 978-0-471-22852-3, 2003.

24. Kehar Singh, Dimple Malik and Naveen Sharma, "Evolving limitations in K-means algorithm in data Mining and their removal", IJCEM International Journal of Computational Engineering & Management, Vol. 12, April 2011.

25. Khamparia, A, Singh, PK, Rani, P, Samanta, D, Khanna, A, Bhushan, B. An internet of health things-driven deep learning framework for detection and classification of skin cancer using transfer learning. Trans Emerging Tel Tech. 2020;e3963. doi:https://doi.org/10.1002/ett.3963

26. Long, Z. Z., Xu, G., Du, J., Zhu, H., Yan, T., & Yu, Y. F. (2021). Flexible Subspace Clustering: A Joint Feature Selection and K-Means Clustering Framework. Big Data Research, 23, 100170.

27. Lopamudra Dey, Sanjay Chakraborty, Anirban Mukhopadhyay. Machine Learning Techniques for Sequence-based Prediction of Viral-Host Interactions between SARS-CoV-2 and Human Proteins. Biomedical Journal, Elsevier, 2020.

28. Martin Ester, Hans-Peter Kriegel, Jorg Sander, Michael Wimmer, Xiaowei Xu, "Incremental clustering for mining in a data ware housing", 24th VLDB Conference New York, USA, 1998.

29. Michael K. Ng, Mark Junjie Li, Joshua Zhexue Huang, and Zengyou He, "On the Impact of Dissimilarity Measure in k-Modes Clustering Algorithm", IEEE transaction on pattern analysis and machine intelligence, vol. 29, No. 3, March 2007.

30. Naresh Kumar Nagwani and Ashok Bhansali, "An Object Oriented Email Clustering Model Using Weighted Similarities between Emails Attributes", International Journal of Research and Reviews in Computer science (IJRRCS), Vol. 1, No. 2, June 2010.

31. Oyelade, O. J, Oladipupo, O. O, Obagbuwa, I. C, "Application of k-means Clustering algorithm for prediction of Students' Academic Performance", (IJCSIS) International Journal of Computer Science and Information security, Vol. 7, No. 1, 2010.

32. Rohan Kumar, Rajat Kumar, Pinki Kumar, Vishal Kumar, Sanjay Chakraborty, Prediction of Protein-Protein interaction as Carcinogenic using Deep Learning Techniques, 2nd International Conference on Intelligent Computing, Information and Control Systems (ICICCS), Springer, pp. 461–475, 2021.

33. Sauravjyoti Sarmah, Dhruba K. Bhattacharyya, "An Effective Technique for Clustering Incremental Gene Expression data", IJCSI International Journal of Computer Science Issues, Vol. 7, Issue 3, No 3, May 2010.

34. Steven Young, Itemer Arel, Thomas P. Karnowski, Derek Rose, University of Tennesee, "A Fast and Stable incremental clustering Algorithm", TN 37996, 7th International 2010.

35. Taoying Li and Yan Chen, "Fuzzy K-means Incremental Clustering Based on K-Center and Vector Quantization", Journal of computers, vol. 5, No. 11, November 2010.

36. Tapas Kanungo, David M. Mount, "An Efficient k-Means Clustering Algorithm: Analysis and implementation," IEEE transaction vol. 24 No. 7, July 2002.

37. Tavallali, P., Tavallali, P., & Singhal, M. (2021). K-means tree: an optimal clustering tree for unsupervised learning. The Journal of Supercomputing, 77(5), 5239–5266.

38. Weka, Waikato environment for knowledge environment - http://www.cs.waikato.ac.nz/ml/weka/.

39. Xiaoke Su, Yang Lan, Renxia Wan, and Yuming, "A Fast Incremental Clustering Algorithm", international Symposium on Information Processing (ISIP'09), Huangshan, P.R. China, August-21-23, pp: 175–178, 2009.

40. Patra B.K., Ville O., Launonen R., Nandi S., Babu K.S. (2013) Distance based Incremental Clustering for Mining Clusters of Arbitrary Shapes. In: Maji P., Ghosh A., Murty M.N., Ghosh K., Pal S.K. (eds) Pattern Recognition and Machine Intelligence. PReMI 2013. Lecture Notes in Computer Science, vol 8251. Springer, Berlin, Heidelberg. doi:https://doi.org/10.1007/978-3-642-45062-4_31.

41. Halkidi M., Spiliopoulou M., Pavlou A. (2012) A Semi-supervised Incremental Clustering Algorithm for Streaming Data. In: Tan PN., Chawla S., Ho C.K., Bailey J. (eds) Advances in Knowledge Discovery and Data Mining. PAKDD 2012. Lecture Notes in Computer Science, vol 7301. Springer, Berlin, Heidelberg. doi:https://doi.org/10.1007/978-3-642-30217-6_48.

42. Zuriana Abu Bakar, Mustafa Mat Deris and Arifah Che Alhadi, "Performance analysis of partitional and incremental clustering", SNATI, ISBN-979-756-061—6, 2005.

43. Chakraborty S., Nagwani N.K. (2011) Analysis and Study of Incremental K-Means Clustering Algorithm. In: Mantri A., Nandi S., Kumar G., Kumar S. (eds) High Performance Architecture and Grid Computing. HPAGC 2011. Communications in Computer and Information Science, vol 169. Springer, Berlin, Heidelberg. doi:https://doi.org/10.1007/978-3-642-22577-2_46

44. Lin J., Vlachos M., Keogh E., Gunopulos D. (2004) Iterative Incremental Clustering of Time Series. In: Bertino E. et al. (eds) Advances in Database Technology - EDBT 2004. EDBT 2004. Lecture Notes in Computer Science, vol 2992. Springer, Berlin, Heidelberg. doi:https://doi.org/10.1007/978-3-540-24741-8_8.

45. A. M. Bagirov, Karmitsa N., Taheri S. (2020) Incremental Clustering Algorithms. In: Partitional Clustering via Nonsmooth Optimization. Unsupervised and Semi-Supervised Learning. Springer, Cham. doi:https://doi.org/10.1007/978-3-030-37826-4_7

46. Joo K.H., Lee W.S. (2005) An Incremental Document Clustering for the Large Document Database. In: Lee G.G., Yamada A., Meng H., Myaeng S.H. (eds) Information Retrieval

Technology. AIRS 2005. Lecture Notes in Computer Science, vol 3689. Springer, Berlin, Heidelberg. doi:https://doi.org/10.1007/11562382_29.

47. Yu H., Zhang C., Hu F. (2014) An Incremental Clustering Approach Based on Three-Way Decisions. In: Cornelis C., Kryszkiewicz M., Ślęzak D., Ruiz E.M., Bello R., Shang L. (eds) Rough Sets and Current Trends in Computing. RSCTC 2014. Lecture Notes in Computer Science, vol 8536. Springer, Cham. doi:https://doi.org/10.1007/978-3-319-08644-6_16.

48. Li Z., Lee JG., Li X., Han J. (2010) Incremental Clustering for Trajectories. In: Kitagawa H., Ishikawa Y., Li Q., Watanabe C. (eds) Database Systems for Advanced Applications. DASFAA 2010. Lecture Notes in Computer Science, vol 5982. Springer, Berlin, Heidelberg. doi:https://doi.org/10.1007/978-3-642-12098-5_3.

Chapter 6
Real-Time Application with Data Mining and Machine Learning

6.1 Introduction

Data mining (DM) is utilised in a variety of fields. Even though there is a number of industrial DM systems on the market today, there are numerous problems in this industry. This chapter will explore the purposes and current research trends of DM and machine learning models.

6.1.1 Data Analysis in Finance

Financial data in banking and finance is often dependable, making systematic data analysis and DM easier [1]. The following are some common scenarios:

- Data warehouse design and construction for different application in data analysis and mining.
- Customer classification and clustering for marketing purposes.
- Money laundering and other financial crimes must be detected and prosecuted.

6.1.2 Industry of Retail

DM has several uses in the retail industry because it collects a large amount of data from sales, client purchase history, product transportation, consumption, and services. Because of the Internet's increasing accessibility, affordability and popularity, it's only natural that the amount of data collected will continue to rise. DM aids in the detection of client buying habits and trends in the retail industry, resulting in

© The Author(s), under exclusive license to Springer Nature Switzerland AG 2022
S. Chakraborty et al., *Data Classification and Incremental Clustering in Data Mining and Machine Learning*, EAI/Springer Innovations in Communication and Computing, https://doi.org/10.1007/978-3-030-93088-2_6

better customer service and more customer retention and satisfaction [2, 3]. The instances of DM in the retail industry are listed below. The benefits of DM are used to design and build data warehouses. On a multidimensional basis, sales, customers, items, time and region are all analysed [4].

6.1.3   Telecommunications Sector

Telecommunications production is one of the fastest-growing sectors today, with provisions such as fax, pager, cell, photos, and more. The telecommunications business continues to grow [5]. This is why DM has become an essential tool for supporting and comprehending business. In the telecommunications manufacturing, DM aids in detecting telecommunication patterns, the exposure of fraudulent actions, better resource use and service quality enhancement. Here are some examples of how DM could be beneficial to telecoms services:

- Analysis of multidimensional associations and sequential patterns
- Telecommunications services for mobile devices
- Analysis of telecommunication data using visualization tools

6.1.4   Analysing Biological Data

In the realm of biology, including genomics, proteomics, functional genomics, and medicinal research, we have seen significant growth in recent years [6]. Bioinformatics includes biological DM as a significant component. The following are some of the ways that DM can help with biological data analysis:

- Integration of disparate, dispersed genomic and proteomic resources using semantics
- Multiple nucleotide sequence alignment, indexing, similarity search and comparative analysis
- Path and association analysis
- Genetic data analysis visualisation tools

6.1.5   Additional Scientific Uses

The statistical techniques are appropriate for handling relatively small and homogeneous datasets in the applications outlined above [7]. A vast amount of information has been gathered from scientific fields such as geosciences, astronomy and others. Rapid numerical simulations in diverse domains fluid dynamics, and so on, many datasets are generated. The uses of DM in the field of scientific applications are listed below.

- Data warehouses and preparation of data
- Mining based on graphs
- Domain-specific knowledge and visualisation

6.2 Detection of Intruders

Intrusion is defined as any action that jeopardizes the integrity, confidentiality or availability of network resources. In today's linked society, security has become a significant concern. Because of the increased use of the Internet and the availability of tools and strategies for breaking into and attacking networks, intrusion detection has become an important part of network administration [8, 9]. The following is a list of places where DM technologies can be used to detect intrusions:

- DM technique for intrusion detection development.
- Association and correlation analysis, as well as aggregation, to aid in the selection and development of distinguishing qualities.
- Stream data analysis.
- DM on a large scale.
- Tools for visualisation and query.

6.2.1 Choosing a DM Methodology

The following characteristics influence the choice of a DM system:

The information could be in ASCII text, a relational database, or a data warehouse. As a result, we must determine what format the DM system can handle. A DM system's interoperability with various operating systems must be considered. A DM system might run on a single operating system or many operating systems [10]. DM systems with web-based user interfaces and XML data input are also available. The data formats in which the DM system will function are referred to as data sources. Some DM systems exclusively deal with ASCII text files, while others work with a variety of relational data sources. For ODBC connections, the DM system should also support OLE DB or ODBC. Integrating DM with databases or data warehouses, DM technologies must be integrated with a database or data warehouse. The linked components are brought together in a unified data processing environment. In scalability in data mining, there are two scalability difficulties [11, 12].

6.3   Applications of Machine Learning

Machine learning plays a vital role in face detection and image recognition tasks. It is mainly responsible for identifying objects (static/moving), persons, places, etc. Facebook uses this feature for user face identification in the profile image. It is a part of the "Deep Face" project in Facebook [13].

Another popular application is computer speech recognition using machine learning methods. Google uses this feature to do searching by voice. The user's voice is first translated into text instructions and then performs the instructed tasks. Google Assistant, Alexa, Cortana, etc., are a kind of devices having this technology [14].

Machine learning is associated with Google Maps to show the correct path and the shortest route from source to destination. Also, it predicts predicts the traffic condition (clear, slow-moving, heavily congested) based on the searched locations. Google or Tesla automatic self-driving cars following the unsupervised learning technique to train the car models of detecting traffic conditions [15].

Recommendation systems are very popular nowadays. Machine learning also plays a key role in such kinds of systems. Different e-commerce companies, like Amazon, E-bay, Flipkart, etc., use this concept for automatic product recommendation to the users, whereas digital entertainment media, such as Hotstar, Netflix, etc., use a similar technique for movies suggestion in the entertainment industry [16].

Machine learning algorithms (decision tree, multilayer perceptron, Naïve Bayes, etc.) provide an essential service while filtering the incoming emails as normal, spam, or important. Some popular spam filters used by Gmail are content filter, header filter, permission filter, etc. [17].

Machine learning helps to detect fraudulent activities during online transactions. There are different ways fraudulent activities can be done during online money transactions. Feedforward neural network helps to find this kinds of fraud transactions [18].

Machine learning also plays an important role for stock market prediction and weather forecasting. Regression and deep learning techniques are extensively used to perform this kind of stock market or predictions [19].

In the field of medical diagnosis, machine learning is used for different disease diagnoses and medical images analysis of patients. With the help of machine learning and deep learning techniques, it is easy to identify the exact position of lesions in the brain and whether a tumour is malignant or benign; whether a patient has diabetes or not can also be detected with the help of supervised machine learning models. Even COVID-19 detection can also be possible in this situation [20].

6.4   Incremental Clustering

This book mainly used two clustering algorithms, K-means and DBSCAN, are selected for incremental implementation. This chapter implements the incremental concept of these two clustering algorithms and compares their characteristics. The incremental clustering technique is used to overcome some of the problems related to the existing K-means and DBSCAN clustering in a dynamic environment. The actual K-means and DBSCAN are the two most popular clustering techniques, but they are not suitable in a dynamic environment where changes happen frequently in databases [21, 22]. The typical clustering algorithms take a lot of time while they are working with dynamic databases. However, from the research point of view, the main goal of this book is to develop and implement two new, improved incremental K-means and DBSCAN clustering algorithms, which can perform suitable clustering of incremental data in a dynamic environment up to some certain point of change in the original database [23, 24]. It will make those algorithms more time-efficient and cost-efficient up to some certain threshold value in the database [43, 44]. In this work, the performance of both the K-means and DBSCAN clustering algorithms is evaluated through basic programming tools and concepts. At last, their performances are also compared with each other to prove which algorithm provides better results on the incremental database. The future scope of the work could be analyzing the other benchmark clustering (hierarchical clustering, grid-based clustering, constraint-based clustering, etc.) techniques in an incremental fashion [25, 26].

6.5   Supervised Learning

6.5.1   Input Dataset Description

The *Pima Indians Diabetes* dataset is considered here as an input. It is one of the widely used datasets collected from Kaggle. It is actually prepared by the National Institute of Diabetes and Digestive and Kidney Diseases. This dataset consists of eight attributes and one target class label representing whether a patient has diabetes (1) or not (0). There is a total of 768 instances. It is a binary classification problem. The main purpose is to predict whether a patient has diabetes or not through diagnostic measurements [27].

Target class variable => Binary: (0 or 1)

6.5.2   Classification Algorithms with Pseudocode on Python Environment

Begin

Input

Import the necessary classification functions from sklearn package. Import load-txt module from numpy package to load our input dataset, and finally load it into our pandas dataframe.

Output

Build a proper classification results. For measuring the performance of an algorithm, we have to consider the following metrics,

"confusion matrix, precision, recall and f-1 score".

Process

1. DB=loadtxt('pima-indians-diabetes').
2. Divide the dataset into input and output variables:

    X=DB[:,0:8]
    Y=DB[:,8]

3. Split the records of DB into training (80%) and testing (20%) sets with validation size = 0.20. We have a total of 768 records. After splitting, the training set will contain 614 records, and the testing set contains 154 records out of 768 records.
4. Now, we train our model with state-of-the-art classification algorithm functions:

    model=KNeighborsClassifier(DB)
    model=GaussianNB(DB)
    model=DecisionTreeClassifier(DB)
    model=SVC(DB, gamma='auto',kernel='poly')
    model=RandomForestClassifier() (DB)

5. Therefore, we append each record with the trained model one by one:

    kNN=KNeighborsClassifier()
    kNN.fit(X, Y)
    GB = GaussianNB()
    GB.fit(X, Y)
    svm = SVC(gamma='auto',kernel='poly')
    #select kernel as per requirements
    svm.fit(X, Y)
    DCT = DecisionTreeClassifier()
    DCT.fit(X, Y)
    rf = RandomForestClassifier()
    rf.fit(X, Y)

6. Now, compute the prediction on the training dataset, find the confusion matrix and calculate the accuracy of the training model.
7. Therefore, fit that model to the testing data if training model accuracy is well accepted.
8. Calculate and print the classification report along with the accuracy, precision, recall and f_1-score from the confusion matrix.

End

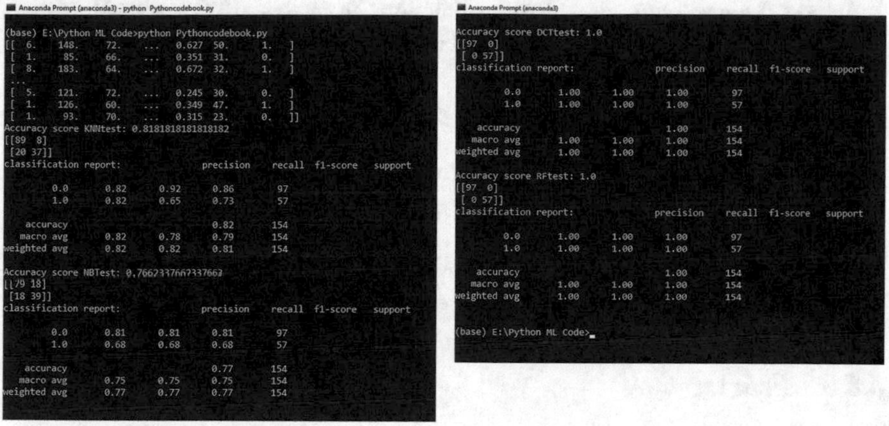

Fig. 6.1 Output screen sort

6.5.3   Output

The results show that our KNN algorithm achieves 81% accuracy (approx.) on the testing set, which is quite impressive. KNN is always not suitable for high dimensionality or categorical features [28, 29]. Similarly, the output for all the other popular classifiers is shown below. The NB algorithm achieves 76.6% accuracy as well, but its performance is not as good enough as KNN. The result shows that our decision tree and random forest algorithms achieve 100% accuracy, the best accuracy compare to all the other classifiers. Figure 6.1 shows output screen sort.

6.6   Classification on Iris Dataset in R Platform

6.6.1   Input Dataset Description

In this framework, the input *iris.csv* dataset is considered. It is one of the popular datasets taken from the UCI machine learning repository. The dataset contains three classes of fifty rows each, where each class represents a kind of iris plant. This dataset consists of four attributes along with one target class.

6.6.2   Pseudocode

Begin

Input

Import the necessary classification functions from the e1071 package. Import loadtxt module from NumPy package to load our input dataset, and finally load it into our pandas dataframe.

Output

Build proper classification results. The final step is to make predictions on our test data [30, 31]. For measuring the presentation of an algorithm, we have to consider the following metrics:

"confusion matrix, precision, recall and f-1 score".

6.6.3   Process

1. DB=load("Iris.csv").
2. First, we delete the species column of the DB due to its string value.

   iris$Species=NULL

3. Split the records of DB into training (80%) and testing (20%) sets with validation size = 0.20. We have total 768 records. After splitting, the training set will contain 614 records, and the testing set contains 154 records out of 768 records.
4. Now, we train our model with state-of-the-art classification algorithm functions:

   myknn<-knn(training, testing, k = 3)
   mysvm<-svm(Species~.,data=iris)
   myNB<- naiveBayes(iris, iris$Species, Species="virginica")
   myrf<- randomForest(Species~., data=iris_training)
   mydectree<- rpart(Species~., data=train, method = 'class')

5. Therefore, we append each record with the trained model one by one:

   Pred1<- predict(myrf, iris_training)
   Pred1<- predict(myknn, iris_training)
   Pred1<- predict(mysvm, iris_training)
   Pred1<- predict(myNB, iris_training)
   Pred1<-predict(mydectree, iris_training)

6. Then, evaluate prediction on the training dataset, compute the confusion matrix and find the accuracy of the training model.

   confusionMatrix(pred1, iris_training$Species)

⇨ Apply to all models

7. Now, fit that model to the testing data if training model accuracy is well accepted.

   Pred2<- predict(myrf, iris_testing)
   Pred2<- predict(myknn, iris_testing)
   Pred2<- predict(mysvm, iris_testing)
   Pred2<- predict(myNB, iris_testing)
   Pred2<- predict(mydectree, iris_testing)

8. Calculate and print the classification report along with the accuracy, precision, recall and f_1-score from the confusion matrix.

   confusionMatrix(pred2, iris_testing$Species)

⇨ Apply to all models

End

6.6.4   Output

```
myknnsetosa versicolor virginica
setosa          7            3           9
versicolor      8           12           6
virginica      10           10          10
```

mysvm

```
setosa versicolor virginica
setosa         50            0           0
versicolor      0           48           2
virginica       0            2          48
```

```
#Misclassification Error: 0.02666667
```

```
MyNB:
Actual
Predicted     setosa versicolor virginica
setosa          50            0           0
versicolor       0           50           0
virginica        0            0          50
```

```
myrf:
Prediction with training data:

Prediction    setosa versicolor virginica
setosa          25            0           0
versicolor       0           25           0
virginica        0            0          25
```

   ⇨ Accuracy: 1.000
   ⇨ CI:0.952
   ⇨ Kappa:1
   ⇨ P-Value:<2.2e-16

Prediction with testing data:

```
Prediction    setosa versicolor virginica
setosa          25            0           0
versicolor       0           23           2
virginica        0            2          23
```

   ⇨ Accuracy:0.9467
   ⇨ CI:0.869
   ⇨ Kappa:0.92
   ⇨ P-Value:<2.2e-16

mydectree:

```
Prediction with testing data:

Prediction    setosa versicolor virginica
setosa         13          0         0
versicolor      0         12         3
virginica       0          1        10
```

> ⇨Accuracy: 0.8974
> ⇨CI:0.7578
> ⇨Kappa:0.8462
> ⇨P-Value:3.435e-13

6.7　Unsupervised Learning

6.7.1　Partitional Clustering with K-Means Clustering

Fig. 6.2 shows Scatter plot of partitional clustering with K-means clustering.

An iris dataset encloses three classes and fifty records each. Each class represents a kind of iris plant. It has four attributes: sepal length, sepal width, petal length and petal width and one binary target class species [32]. Figure 6.3 shows graph for partitional clustering with K-means clustering A; scatter plot of the dataset is shown in the below Fig. 6.2.

DB=iris

Output

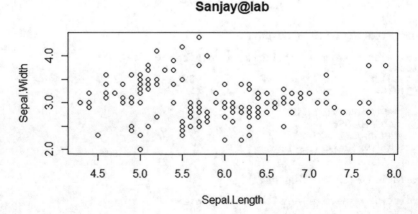

Fig. 6.2 Scatter plot of partitional clustering with K-means clustering

An output with three different clusters to differentiate those three species (setosa, versicolor and virginica).

6.7.2   Pseudocode of K-Means (DB,k)

Assume the initial number of clusters (k) = 3.
Plot those centroids randomly.

Repeat until convergence or a certain number of fixed iterations:
For each data point:

– Calculate the nearest centres using a suitable distance metric on DB. *function(k,iris,3)*
– Assign the point to the cluster with nearest distance.

For each cluster up to k:

– Evaluate new mean of all data points to each cluster.
– Replace the new means with the old centroids to each cluster.

End
Now, K-means clustering is called with three sizes of clusters 38, 50 and 62.
Figure 6.3 shows graph for partitional clustering with K-means clustering.

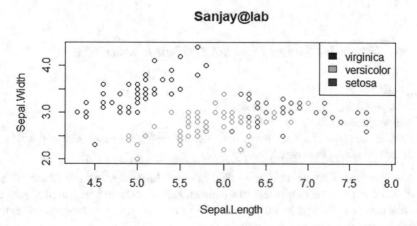

Fig. 6.3 Graph for partitional clustering with K-means clustering

6.7.3   Cluster Means

Sepal.LengthSepal.WidthPetal.LengthPetal.Width

	Sepal.Length	Sepal.Width	Petal.Length	Petal.Width
1	6.850000	3.073684	5.742105	2.071053
2	5.006000	3.428000	1.462000	0.246000
3	5.901613	2.748387	4.393548	1.433871

[1] 2 2 2 2 2 2 2 2 2 2 2 2 2 2 2 2 2 2 2 2 2 2 2 2 2 2 2 2 2 2 2 2 2 2 2 2 2 2 2 2

[41] 2 2 2 2 2 2 2 2 2 2 3 3 1 3 3 3 3 3 3 3 3 3 3 3 3 3 3 3 3 3 3 3 3 3 3 3 3 1 3 3

[81] 3 3 3 3 3 3 3 3 3 3 3 3 3 3 3 3 3 3 3 3 1 3 1 1 1 3 1 1 1 1 1 3 3 1 1 1 1 3

[121] 1 3 1 3 1 1 3 3 1 1 1 1 1 3 1 1 1 1 3 1 1 1 3 1 1 1 3 1 1 1 3 1 1 3

cluster sum of squares:

[1] 23.87947 15.15100 39.82097

(between_SS / total_SS= 88.4 %)

setosaversicolor virginica

	setosa	versicolor	virginica
1	0	2	36
2	50	0	0
3	0	48	14

6.7.4   Elbow Method to Find the Optimum Value of K

K_{max}<-20
w<- sapply(2:K_{max}, function(k,iris,3))
elb<-data.frame(2:K_{max}, w)
ggplot(elb, aes(x = X2.K_{max}, y = wss)) +geom_point() +geom_line() + scale_x_
continuous(breaks = seq(1, 20, by = 1))

Now, instead of taking the initial assumption of the k value, we use elbow method to seek the optimal set of clusters in a dataset. Elbow method is required to find the optimal number of initial k clusters and rerun the K-means clustering with the optimal k. Figure 6.4 shows elbow method.

Initial Output (k=5)

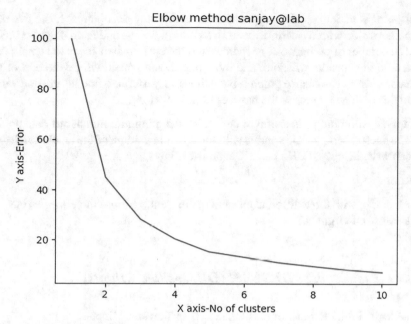

Fig. 6.4 Elbow method

[0 0 0 3 3 0 0 0 0 3 0 0 0 3 3 3 3 3 3 3 3 0 0 0 0 0 3 3 0 0 3 2 1 2 4 1 4

1 0 1 4 4 1 4 1 4 1 4 4 4 4 1 1 1 1 1 1 1 1 1 4 4 4 4 4 1 4 2 1 1 2 0 2 1

2 1 1 1 4 4 1 1 2 2 4 2 4 2 1 1 2 1 1 1 2 2 2 1 1]

From the above figure, it is clearly seen that the optimal value of k exists in between 2 and 4, as the elbow-like shape is formed at k=3 in the above graph. However, the updated output of three respective clusters is

[1 1 1 1 1 1 1 1 1 1 1 1 1 1 1 1 1 1 1 1 1 1 1 1 1 1 1 1 1 1 1 1 1 1 0 0 0 2 0 2

0 2 0 2 2 2 2 2 2 0 2 2 2 2 2 2 2 2 0 0 0 0 2 2 2 2 2 2 0 2 0 0 0 0 2 0 0

0 0 0 0 2 2 0 0 0 0 2 0 2 0 2 0 0 2 2 0 0 0 0 0 2]

6.8 Density-Based Clustering

Input

An iris dataset has three classes and fifty records each. Each class represents a type of iris plant. It has four attributes: sepal length, sepal width, petal length, and petal width, and one binary target class species DB = iris.

epsilon: It is mainly used to represent two neighboring data points. If the two data points reside within a certain distance (epsilon), we can call them neighbors [33]. The number of outliers can be increased if the epsilon distance is too small; otherwise the clusters will join, and most of the data points will be the part of the same cluster. K-distance graph is sometimes very useful to find the epsilon value. In this case, we initialise the basic epsilon value [34].

Minpts: epsilon radius is useful to determine the minimum number of data points. The chosen Minpts value varies with the size and volume of the database. So, we can write Minpts $\geq$ DB$_{dimension}$. Generally, Minpts$_{min}$ = 3.

Output

An output with three different clusters to differentiate those three species (setosa, versicolor and virginica).

6.8.1 Pseudocode: DBSCAN (DB, epsilon, Minpts)

Assume the initial number of clusters (k)=3, epsilon and Minpts.
   Plot those centroids randomly and initialise a cluster index ct=1.
   Repeat until convergence or a certain number of fixed iterations:
   For each unvisited data point x in DB:

- x is being visited first.
- Find the neighbours (N) by calculating the nearest points from x using a suitable distance metric. *function (DBSCAN, iris, epsilon, Minpts)*
- if |N| $\geq$ Minpts& dis(x,N) $\leq$ epsilon

    N = N U N$_1$
    If x$_1$ does not belong any cluster:
    add x$_1$ to cluster ct.

End

6.8.2 Output

We first Consider, dbscan(DB, epsilon = 0.45, minPts = 5)
   DBSCAN contains 2 clusters and 24 noise points. Figures 6.5 and 6.6 show DBSCAN clusters.

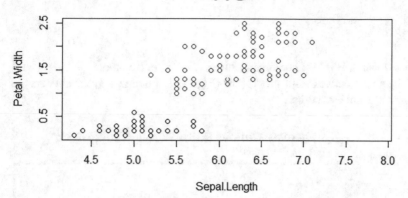

Fig. 6.5 DBSCAN clusters

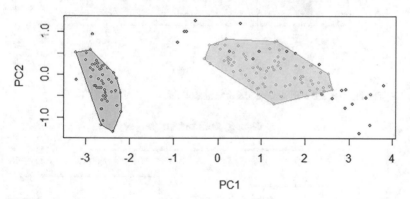

Fig. 6.6 DBSCAN clusters

```
0      1      2
24     48     78
```

DBSCAN Clusters:
[1] 1 1 1 1 1 1 1 1 1 1 1 1 1 1 1 1 1 1 1 1 1 1 1 1 1 1 0 1 1 1 1 1 1 1 1 1 1 1 1 1 1
[41] 1 0 1 1 1 1 1 1 1 1 1 2 2 2 2 2 2 2 2 0 2 2 0 2 0 2 0 2 2 2 2 2 0 2 2 2 2 2 2 2 2 2
[81] 2 2 2 2 2 2 2 0 2 2 2 2 2 0 2 2 2 2 0 2 2 2 2 2 2 0 0 0 0 0 2 2 2 2 0 2 2 0 0 2
[121] 2 2 0 2 2 0 2 2 2 0 0 0 2 2 0 0 2 2 2 2 2 2 2 2 2 2 2 2 2

setosa versicolor virginica
```
   0    2      7     15
   1   48      0      0
   2    0     43     35
```

If we modify the epsilon value to 0.4 and Minpts value to 4, then the output will be

```
0   1   2   3   4
25  47  38  36   4
```

However, DBSCAN contains 4 clusters and 25 noise points.

The above convex hulls plots show (Figs. 6.7, 6.8, and 6.9) the clusters are well separated from each other.

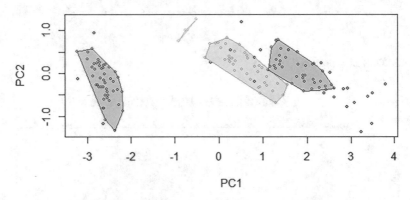

Covex Cluster Hulls_Plot-2_Sanjay@lab

Fig. 6.7 The clusters are well separated from each other

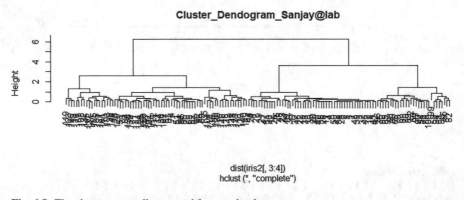

Cluster_Dendogram_Sanjay@lab

dist(iris2[, 3:4])
hclust (*, "complete")

Fig. 6.8 The clusters are well separated from each other

6.9   Hierarchical Clustering Using Agglomerative Clustering

1. Start with n number of clusters. Each cluster contains one data object. Cluster ϵ $\{1,...,n\}$.
2. Calculate the distance between-cluster $Dist(ob1, ob2)$ as the between-object distance of the two objects in $ob1$ and $ob2$, respectively, $ob1, ob2 = 1, 2, ..., n$. Let

Cluster_Dendogram_Sanjay@lab

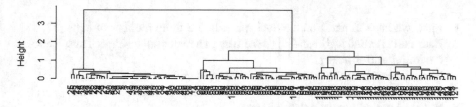

dist(iris2[, 3:4])
hclust (*, "average")

Fig. 6.9 The clusters are well separated from each other

the square matrix $D = (Dist(ob_1, ob_2))$. If the objects are quantitative in nature, then Euclidean distance can be the best distance measure.

Repeat $n-1$ times until there is only one cluster:

3. Then, we select the minimum distance $Dist(ob_1, ob_2)$ among all distances, and find the most similar pair of clusters ob_1 and ob_2.
4. Join *ob1* and *ob2* to form a new cluster *ct* and calculate the between-cluster distance $Dist(ct, ck)$ for any existing cluster $ck \neq ob_1, ob_2$. Now, we can remove the corresponding rows and columns of ob_1 and ob_2 in the distance matrix and modify the cluster ct by adding a new row and column in the matrix [35, 36].

End

6.9.1   Output of Complete Linkage Method

```
clusterCutsetosa versicolor virginica
```

clusterCut	setosa	versicolor	virginica
1	50	0	0
2	0	21	50
3	0	29	0

Mean linkage method

```
clusterCutsetosa versicolor virginica
```

clusterCut	setosa	versicolor	virginica
1	50	0	0
2	0	45	1
3	0	5	49

6.10   Grid-Based Clustering

- First, we have to partition the data space into a finite number of cells [37, 38]. Each cell is called a grid and it is generally in a rectangular shape. Thus, we form the initial grid structure.
- Now, we compute the cell density of each grid cell.
- Then we arrange those cell densities in ascending or descending order.
- We need to find out the cluster centres.
- Now, we start traversing the neighbour cells accordingly.

  Some popular grid-based clustering methods are STING and CLIQUE.

6.10.1   Coding Snapshots of the Experiment

Some important functions of this research work are given below in Java:

(i) **Actual K-means algorithm implementation using Java**

// Reading original file

public actualkmeans(String filename)throws Exception

{

  String[] options;

  Instances data = new Instances(new BufferedReader(new FileReader(filename)));

  System.out.println("\n--> K-means clutering");

// Running simple K-means from Weka

  SimpleKMeans kMeans = new SimpleKMeans();

  kMeans.setNumClusters(5);

  System.out.println(ClusterEvaluation.evaluateClusterer(kMeans,options));

// calculate number of iteration

  kMeans.buildClusterer(data);

  String kmeansresult=kMeans.toString();

  String []strArr=kmeansresult.split("\n");

  String supportLine=strArr[4];

  String []suppV = supportLine.split(":");

    String sup = suppV[1];

    sup = sup.split("\\(")[0];

    support1 = Double.parseDouble(sup);

```
System.out.println(support1);
```

// splitting and storing means

```
String[] result = "321.376238,164.366337,10.128713,92.415842".split(",");

for (int x=0; x<result.length; x++)

System.out.println(result[x]);

System.out.println("\n");
```

// Required time calculation

```
public long timerequirement()throws Exception

   {

      long start;

      long end;

      start=System.currentTimeMillis();

      new actualkmeans("d:\\original air pollution data.arff");

      end=System.currentTimeMillis();

      long Timerequired=(end - start);

      System.out.println("---The needed time for the actual k-means on original air
pollution data.arff file: "+Timerequired+" ms");

      return Timerequired; } }
```

6.11   Explanation of the Typical K-Means Algorithm Code

At first step, the air pollution database (4D data) which is stored in .arff file format is read by the help of "BufferedReader" class and "FileReader" method and stored into data variable [39, 40]. Then the simple K-means algorithm runs on that data after assuming five clusters initially. As a result, five means (each mean have four values) come out and store those resultant means into an array through splitting and parsing methods. At last the total time is calculated for the typical K-means algorithm [41, 42].

// Reading incremental data

```
ArrayList<String> list = showLines("d:\\incremental air pollution data.arff", 7,507);
```

// Covert incremental data into double format

```
double darray[][]=new double[list.size()][4];

      System.out.println("\n");

      System.out.println("New incoming data are converted into double");
```

```java
for (int r = 0; r<list.size(); r++)

{

        for (int c = 0; c<4; c++)

        {

darray[r][c]=Double.parseDouble(twoDArray[r][c]);

        //System.out.print(" Result is:"+darray[r][c]);

                }

        //System.out.print("\n");

}
```

//minimum distance calculation directly from mean values

```java
4. for(int m=0; m<darray.length;m++)

{

    for(int n=0;n<darray[m].length;n++)

            {

    double distance1 = CalculateDistance ( d0, d1, d2, d3,darray[m][n]);

    double distance2 = CalculateDistance ( e0, e1, e2, e3,darray[m][n]);

    double distance3 = CalculateDistance ( f0, f1, f2, f3,darray[m][n]);

    double distance4 = CalculateDistance ( g0, g1, g2, g3,darray[m][n]);

    double distance5 = CalculateDistance ( h0, h1, h2, h3,darray[m][n]);

    double minDist = minimum(distance1,distance2,distance3,distance4,distance5);

  if(minDist == distance1)

        {

    finalResult[m] = 1;

        }

  else if(minDist == distance2)

        {  finalResult[m] = 2;

        }

  else if(minDist == distance3)

        {  finalResult[m] = 3;
```

```
        }
    else if(minDist == distance4)
        {
    finalResult[m] = 4;
        }
    else if(minDist == distance5)
            {
            finalResult[m] = 5;
            }
            System.out.println("The Data is belonging to cluster : " + finalResult[m]);
            }}
```

// **Body of CalculateDistance function**

```
public static double CalculateDistance(double x1, double  y1,double  z1, double k1,double
pass)
{
 double diffx= x1 - pass;
  double diffy= y1 - pass;
  double diffz = z1 - pass;
  double diffk = k1 - pass;
  double diffx_sqr = (diffx*diffx);
  double diffy_sqr = (diffy*diffy);
  double diffz_sqr = (diffz*diffz);
  double diffk_sqr = (diffk*diffk);
  double Fdistance = Math.sqrt(diffx_sqr + diffy_sqr + diffz_sqr + diffk_sqr);
  return Fdistance;
}
```

// **Calculate minimum Euclidean distance**

```
  public static double minimum(double E1,double E2,double E3,double E4,double E5)
    {
```

```java
double min = 0;
        if (E1<E2 && E1<E3 && E1<E4 && E1<E5)
{

        System.out.println("The minimum distance of the current data is:"+E1);
        min = E1;

}
else if(E2<E1 && E2<E3 && E2<E4 && E2<E5)
{

            System.out.println("The minimum distance of the current data is:"+E2);
            min = E2;

    }
else if(E3<E1 && E3<E2 && E3<E4 && E3<E5)
{

            System.out.println("The minimum distance of the current data is:"+E3);
            min = E3;

    }
else if(E4<E1 && E4<E2 && E4<E3 && E4<E5)
{

            System.out.println("The minimum distance of the current data is:"+E4);
            min = E4;

    }
else if(E5<E1 && E5<E2 && E5<E3 && E5<E4)
    {

            System.out.println("The minimum distance of the current data is:"+E5);
            min = E5;

    }
return min;

}
```

6.12   Applications of Classifiers for Protein-Protein Prediction Between SARS-CoV-2 and Human

The input *SARS-CoV-2-human PPI (Protein-Protein Interactions)* dataset is considered here. The PPI database is made between *Homo sapiens* proteins and coronavirus proteins. In this database, 332 unique interactions are found between *Homo sapiens* proteins and 24 coronavirus proteins. These 332 human proteins are considered to make positive training and testing datasets. The feature vectors of the training dataset are amino acid composition, conjoint triad and pseudo-amino acid composition. Then, we have applied some state-of-the-art classifiers for training our model and discovered some new sets of interactions through extensive analysis. After the feature selection, we have reduced multiple features from 413 to 28. The application of "feature selection LVQ" leaves for the reader as a preprocessing step.

6.12.1   Pseudocode (Python Platform)

Begin

Input

Import the necessary classification functions from sklearn package. Import load-txt module from numpy package to load our input PPI dataset, and finally load it into our pandas dataframe.

Output

Build a proper classification results. The final step is to make predictions on our test data. For measuring the performance of an algorithm, we have to consider the following metrics.

Process

DB=loadtxt('*SARS-CoV-2-human PPI*')
Divide the PPI dataset into input and output variables:

> *dataset1=pd.read_csv("featrain.csv");*
> *print(dataset1)*
> *X = dataset1.iloc[:,0:28].values*
> *Y = dataset1.iloc[:,28].values*
> *dataset2=pd.read_csv("featest.csv");*
> *print(dataset2)*
> *X_validation = dataset2.iloc[:,0:28].values*
> *Y_validation= dataset2.iloc[:,28].values*

Split the records of DB into training (80%) and testing (20%) sets with validation size = 0.20. The entire dataset is divided into two class labels where positive represents 1 and negative represents 0. Positive dataset represents the interacting proteins between human and coronavirus, whereas the negative dataset represents non-interacting proteins from HPRD. Now, we train our model with state-of-the-art classification algorithm functions:

```
model=KNeighborsClassifier(DB);
model=GaussianNB(DB);
model=DecisionTreeClassifier(DB);
model= SVC(DB, gamma='auto',kernel='poly');
model=RandomForestClassifier(DB);
model=XGBclassifier(DB);
model=AdaBoostClassifier(n_estimators=50,learning_rate=1);
```

Therefore, we append each record with the prepared model one by one. The training set accuracy is given at the Output section.

```
kNN=KNeighborsClassifier();
kNN.fit(X, Y);
```

```
GB = GaussianNB();
GB.fit(X, Y);
svm = SVC(gamma='auto',kernel='poly');
svm.fit(X, Y);
DCT = DecisionTreeClassifier();
DCT.fit(X, Y);
rf = RandomForestClassifier();
rf.fit(X, Y);
xg=XGBclassifier();
xg.fit(X, Y);
ad=AdaBoostClassifier();
ad.fit(X, Y);
```

Therefore, fit that model to the testing data if training model accuracy is well accepted. Start making predictions on validation dataset using trained classifier models.

```
Predict<- predict(rf, iris_testing);
Predict<- predict(KNN, iris_testing);
Predict<- predict(svm, iris_testing);
Predict<- predict(GB, iris_testing);
Predict<- predict(DCT, iris_testing);
Predict->predict(xg, iris_testing);
Predict->predict(ad, iris_testing);
```

Calculate and print the classification report along with the accuracy, precision, recall and f1-score from the confusion matrix.

```
confusion_matrix(Y_validation, predict);
classification_report(Y_validation, predictions);
```

accuracy_score(Y_validation, predictions);

Apply for all models

End

6.13   Deep Multilayer Perceptron (DMLP) for Protein-Protein Prediction Between SARS-CoV-2 and *Homo sapiens*

6.13.1   *Pseudocode (Python Platform)*

Begin

Input

   Import the necessary classification functions from sklearn and Keras packages. Keras is used to import deep sequential neural network. Import loadtxt module from numpy package to load our input PPI dataset, and finally load it into our pandas dataframe.

Output

   Build a proper classification results. The final step is to make predictions on our test data. For measuring the performance of an algorithm, we have to consider the following metrics"

   "Confusion matrix, accuracy, precision, recall and f-1 score".

Process

1. DB=loadtxt('*SARS-CoV-2-human PPI*');
2. Divide the PPI dataset into input and output variables,

   dataset=pd.read_csv("featrain.csv");
   X = dataset.iloc[:,0:28].values
   y= dataset.iloc[:,28].values
   dataset1=pd.read_csv("featest.csv");
   Z = dataset1.iloc[:,0:28].values
   k= dataset1.iloc[:,28].values

3. Split the records of DB into training (80%) and testing (20%) sets with validation size = 0.20. The entire dataset is divided into two class labels where positive represents 1 and negative represents 0. Positive dataset represents the interacting proteins between human and coronavirus, whereas the negative dataset represents non-interacting proteins from HPRD.
4. Now, we train our model with state-of-the-art classification algorithm functions:

model=Dense(30, input_dim=28, activation='relu');
model=Dense(26, activation='relu');
model=Dense(22, activation='relu');
model=Dense(18, activation='relu');
model=Dense(14, activation='relu');
model=Dense(10, activation='relu');
model=Dense(8, activation='relu');
model=Dense(1, activation='sigmoid');

5. Therefore, compile the Keras model with the help of stochastic gradient descent and binary cross-entropy loss function. Finally fit the model on the training data.

    model.fit(X, y, epochs=50, batch_size=10)

6. Therefore, fit that model to the testing data if training model accuracy is well accepted. Start making predictions on validation dataset using the above DMLP model.
7. Calculate and print the classification report along with the accuracy, precision, recall and f1-score from the confusion matrix.

End

6.14   Conclusion

The most expressive research and application domains are data mining and machine learning. All real-time applications rely on data mining and machine learning, either directly or indirectly. Numerous data mining and machine learning applications have been found, including data analysis in finance, retail, telecommunications, biological data analysis, extra-scientific usage, and intrusion detection. Because it captures a vast amount of data from sales, client purchase history, product transportation, consumption, and services, data mining has various applications in the retail industry. It's only logical that the amount of data collected will continue to rise as the Internet becomes more accessible, affordable, and popular. In the retail industry, data mining aids in the detection of client buying behaviors and trends, leading to improved customer service and increased customer retention and satisfaction.

References

1. Aristidis Likasa, Nikos Vlassis, Jakob J. Verbeek ," The global k-means clustering algorithm", the journal of the pattern recognition society, Pattern Recognition36 (2003) 451–461, 2002.
2. Carlos Ordonez, "Clustering Binary Data Streams with K-means", San Diego, CA, USA. Copyright 2003, ACM 1- 58113-763-x, DMKD'03, June 13, 2003.

3. K. Wang et al., "A Trusted Consensus Scheme for Collaborative Learning in the Edge AI Computing Domain," in IEEE Network, vol. 35, no. 1, pp. 204–210, January/February 2021, doi: https://doi.org/10.1109/MNET.011.2000249.

4. Guha, D. Samanta, A. Banerjee and D. Agarwal, "A Deep Learning Model for Information Loss Prevention From Multi-Page Digital Documents," in IEEE Access, vol. 9, pp. 80451–80465, 2021, doi: https://doi.org/10.1109/ACCESS.2021.3084841.

5. Rohan Kumar, Rajat Kumar, Pinki Kumar, Vishal Kumar, Sanjay Chakraborty, Prediction of Protein-Protein interaction as Carcinogenic using Deep Learning Techniques, 2nd International Conference on Intelligent Computing, Information and Control Systems (ICICCS), Springer, pp. 461–475, 2021.

6. Guha, A., Samanta, D. Hybrid Approach to Document Anomaly Detection: An Application to Facilitate RPA in Title Insurance. Int. J. Autom. Comput. 18, 55–72 (2021). https://doi.org/10.1007/s11633-020-1247-y

7. Lopamudra Dey, Sanjay Chakraborty, Anirban Mukhopadhyay. Machine Learning Techniques for Sequence-based Prediction of Viral-Host Interactions between SARS-CoV-2 and Human Proteins. Biomedical Journal, Elsevier, 2020.

8. Khamparia, A, Singh, PK, Rani, P, Samanta, D, Khanna, A, Bhushan, B. An internet of health things-driven deep learning framework for detection and classification of skin cancer using transfer learning. Trans Emerging Tel Tech. 2020;e3963. https://doi.org/10.1002/ett.3963

9. Jiawei Han and Micheline Kamber, Data Mining concepts and techniques, Morgan Kaufmann (publisher) from chapter-7 'cluster analysis', ISBN:978-1-55860-901-3, 2006.

10. Dunham, M.H., Data Mining: Introductory And Advanced Topics, New Jersey: Prentice Hall, ISBN-13: 9780130888921. 2003.

11. H.Witten, Data mining: practical machine learning tools and techniques with Java implementations San-Francisco, California : Morgan Kaufmann,ISBN: 978-0-12-374856-0 2000.

12. Kantardzic, M. Data Mining: concepts, models, method, and algorithms, New Jersey: IEEE press, ISBN: 978-0-471-22852-3, 2003.

13. Michael K. Ng, Mark Junjie Li, Joshua Zhexue Huang, and Zengyou He, " On the Impact of Dissimilarity Measure in k-Modes Clustering Algorithm ", IEEE transaction on pattern analysis and machine intelligence, vol.29, No. 3, March 2007.

14. Nareshkumar Nagwani and Ashok Bhansali, "An Object Oriented Email Clustering Model Using Weighted Similarities between Emails Attributes", International Journal of Research and Reviews in Computer science (IJRRCS), Vol. 1, No. 2, June 2010.

15. Oyelade, O.J, Oladipupo, O. O, Obagbuwa, I. C, "Application of k-means Clustering algorithm for prediction of Students' Academic Performance",(IJCSIS) International Journal of Computer Science and Information security,Vol.7,No. 1, 2010.

16. S.Jiang, X.Song, "A clustering based method for unsupervised intrusion detections" . Pattern Recognition Letters, PP.802–810, 2006.

17. Steven Young, ItemerArel, Thomas P. Karnowski,Derek Rose, University of Tennesee, "A Fast and Stable incremental clustering Algorithm", TN 37996, 7th International 2010.

18. Taoying Li and Yan Chen, "Fuzzy K-means Incremental Clustering Based on K-Center and Vector Quantization", Journal of computers, vol. 5, No.11, November 2010.

19. Tapas Kanungo , David M. Mount , "An Efficient k-Means Clustering Algorithm:Analysis and implementation IEEE transaction vol. 24 No. 7, July 2002.

20. Zuriana Abu Bakar, Mustafa Mat Deris and ArifahCheAlhadi, "Performance analysis of partitional and incremental clustering", SNATI, ISBN-979-756-061-6, 2005.

21. Xiaoke Su, Yang Lan, Renxia Wan, and Yuming, "A Fast Incremental Clustering Algorithm ", international Symposium on Information Processing (ISIP'09), Huangshan, P.R.China, August-21–23, pp:175–178,2009.

22. Kehar Singh, Dimple Malik and Naveen Sharma, "Evolving limitations in K-means algorithm in data Mining and their removal", IJCEM International Journal of Computational Engineering & Management, Vol. 12, April 2011.

23. Anil Kumar Tiwari, Lokesh Kumar Sharma, G. Rama Krishna, " Entropy Weighting Genetic k-Means Algorithm for Subspace Clustering ",International Journal of Computer Applications (0975–8887),Volume 7– No.7, October 2010.

24. K. Mumtaz, Dr. K. Duraiswamy, "An Analysis on Density Based Clustering of Multi Dimensional Spatial Data", Indian Journal of Computer Science and Engineering, Vol. 1 No 1, pp-8-12, ISSN : 0976-5166.

25. A.M.Sowjanya, M.Shashi, "Cluster Feature-Based Incremental Clustering Approach (CFICA) For Numerical Data, IJCSNS International Journal of Computer Science and Network Security, VOL.10 No.9, September 2010.

26. Martin Ester, Hans-Peter Kriegel, Jorg Sander, MichaelWimmer, Xiaowei Xu, "Incremental clustering for mining in a data ware housing", 24th VLDB Conference New York, USA, 1998.

27. Sauravjyoti Sarmah, Dhruba K. Bhattacharyya,"An Effective Technique for Clustering Incremental Gene Expression data" , IJCSI International Journal of Computer Science Issues, Vol. 7, Issue 3, No 3, May 2010.

28. Debashis Das Chakladar and Sanjay Chakraborty, Multi-target way of cursor movement in brain computer interface using unsupervised learning, Biologically Inspired Cognitive Architectures (Cognitive Systems Research), Elsevier, 2018.

29. Althar, R.R., Samanta, D. The realist approach for evaluation of computational intelligence in software engineering. Innovations Syst Softw Eng 17, 17–27 (2021). https://doi.org/10.1007/s11334-020-00383-2.

30. B. Naik, M. S. Obaidat, J. Nayak, D. Pelusi, P. Vijayakumar and S. H. Islam, "Intelligent Secure Ecosystem Based on Metaheuristic and Functional Link Neural Network for Edge of Things," in IEEE Transactions on Industrial Informatics, vol. 16, no. 3, pp. 1947–1956, March 2020, doi: https://doi.org/10.1109/TII.2019.2920831.

31. Debashis Das Chakladar and Sanjay Chakraborty, EEG Based Emotion Classification using Correlation Based Subset Selection, Biologically Inspired Cognitive Architectures (Cognitive Systems Research), Elsevier, 2018.

32. D. Samanta et al., "Cipher Block Chaining Support Vector Machine for Secured Decentralized Cloud Enabled Intelligent IoT Architecture," in IEEE Access, vol. 9, pp. 98013–98025, 2021, doi: https://doi.org/10.1109/ACCESS.2021.3095297.

33. CHEN Ning , CHEN An, ZHOU Long-xiang, "An Incremental Grid Density-Based Clustering Algorithm", Journal of Software, Vol.13, No.1,2002.

34. Bock, Frederic E., et al. "A Review of the Application of Machine Learning and Data Mining Approaches in Continuum Materials Mechanics." Frontiers in Materials, vol. 6, 2019, p. 110. Frontiers, https://doi.org/10.3389/fmats.2019.00110.

35. Amador, Sandra, et al. "Chapter 6 – Data Mining and Machine Learning Techniques for Early Detection in Autism Spectrum Disorder." Neural Engineering Techniques for Autism Spectrum Disorder, edited by Ayman S. El-Baz and Jasjit S. Suri, Academic Press, 2021, pp. 77–125. ScienceDirect, https://doi.org/10.1016/B978-0-12-822822-7.00006-5.

36. Dabhade, Pranav, et al. "Educational Data Mining for Predicting Students' Academic Performance Using Machine Learning Algorithms." Materials Today: Proceedings, June 2021. ScienceDirect, https://doi.org/10.1016/j.matpr.2021.05.646.

37. Dogan, Alican, and Derya Birant. "Machine Learning and Data Mining in Manufacturing." Expert Systems with Applications, vol. 166, Mar. 2021, p. 114060. ScienceDirect, https://doi.org/10.1016/j.eswa.2020.114060.

38. Emami Javanmard, Majid, et al. "Data Mining with 12 Machine Learning Algorithms for Predict Costs and Carbon Dioxide Emission in Integrated Energy-Water Optimization Model in Buildings." Energy Conversion and Management, vol. 238, June 2021, p. 114153. ScienceDirect, https://doi.org/10.1016/j.enconman.2021.114153.

39. Jimenez-Carvelo, Ana M., and Luis Cuadros-Rodríguez. "Data Mining/Machine Learning Methods in Foodomics." Current Opinion in Food Science, vol. 37, Feb. 2021, pp. 76–82. ScienceDirect, https://doi.org/10.1016/j.cofs.2020.09.008.

40. Lord, Dominique, et al. "Chapter 12 – Data Mining and Machine Learning Techniques." Highway Safety Analytics and Modeling, edited by Dominique Lord et al., Elsevier, 2021, pp. 399–428. ScienceDirect, https://doi.org/10.1016/B978-0-12-816818-9.00016-0.

41. Ma, Ying, et al. "Meta-Analysis of Cellular Toxicity for Graphene via Data-Mining the Literature and Machine Learning." Science of The Total Environment, vol. 793, Nov. 2021, p. 148532. ScienceDirect, https://doi.org/10.1016/j.scitotenv.2021.148532.

42. Yang, Xin-She. "Chapter 16 – Data Mining and Deep Learning." Nature-Inspired Optimization Algorithms (Second Edition), edited by Xin-She Yang, Academic Press, 2021, pp. 239–58. ScienceDirect, https://doi.org/10.1016/B978-0-12-821986-7.00023-8.

43. Zhao, Qingkun, et al. "Machine Learning-Assisted Discovery of Strong and Conductive Cu Alloys: Data Mining from Discarded Experiments and Physical Features." Materials & Design, vol. 197, Jan. 2021, p. 109248. ScienceDirect, https://doi.org/10.1016/j.matdes.2020.109248.

44. Zou, Chengxiong, et al. "Integrating Data Mining and Machine Learning to Discover High-Strength Ductile Titanium Alloys." Acta Materialia, vol. 202, Jan. 2021, pp. 211–21. ScienceDirect, https://doi.org/10.1016/j.actamat.2020.10.056.

Chapter 7
Feature Subset Selection Techniques with Machine Learning

7.1 Introduction

With the fast advancement of cutting-edge technologies, newer massive computer and Internet solutions have produced massive quantity of data at an excessive rate, including video, image, text, audio, and data derived from societal relationships, and the emergence of the IoT and cloud computation. These data often contain many dimensions, which makes data analysis and decision-making difficult. In both theory and practice, variable selection has been shown to be useful in analyzing high-dimensional data and improving learning capability. The mechanism of getting a variant from an initial variant set as per variable selection criteria that choose the required characteristics of the dataset is called variable selection. It helps in the compression of data processing scales by eliminating superfluous and unnecessary information. Learning algorithms may be preprocessed using variable selection techniques, and excellent variable selection outcomes may increase learning accuracy, shorten learning time duration and minimize learning outcomes [1]. Two methods for dimensional reduction are variable selection and variant extraction. Unlike variable selection, variant extraction often requires transforming actual data into variants with high pattern recognition capacity, while the actual data may be considered variants with low recognition capability. Image processing, image retrieval, intrusion detection system, bioinformatics data analysis fault diagnostics and other areas utilize variable selection, a study subject in methodology and practice for years. Variable selection techniques depend on statistics, information theory, manifold, and rough set and may be classified based on different criteria, depending on the theoretical concept. (a) Variable selection techniques may be classified as supervised, unsupervised, or semi-supervised depending on the kind of training data used (labeled, unlabeled, or partly labeled). This research depicts a complete architecture for supervised, unsupervised, and semi-supervised variable selection. (b) Variable selection techniques are categorized as filter, wrapper or embedding

© The Author(s), under exclusive license to Springer Nature Switzerland AG 2022
S. Chakraborty et al., *Data Classification and Incremental Clustering in Data Mining and Machine Learning*, EAI/Springer Innovations in Communication and Computing, https://doi.org/10.1007/978-3-030-93088-2_7

models based on their connection with learning techniques. (c) Based on their search methodologies, variable selection techniques may be classified into forward increase, backward deletion, random and hybrid frameworks. (d) Variable selection techniques may be split into variant rank, and variant choice approaches depending on the kind of output. Only the connection of the variant with the class label is taken into account by the filter approach. It has a lower computing cost in comparison to the wrapper model. The filter model's assessment criteria are crucial. Meanwhile, the embedded model chooses variants during the learning model's training process, and the variable selection result is outputted automatically after the training process is completed. While the total of the absolute readings in regression coefficients is lower than a constant, Lasso minimizes the sum of squares of residuals, yielding stringent regression coefficients become zero. The variables are then truncated using the AIC and BIC criteria, and the dimension reduction is achieved. Lasso-dependent variable selection techniques, such as Lasso in a regression model and others offer good stability. For high-dimensional data, Lasso approaches are subject to enormous computational cost or overfitting issues [2, 3]. The machine learning model is typically used to assess the usefulness of the variable selection technique. A good variable selection technique must have higher learning accuracy while using little computing resources. Though there are thorough studies of variable selection, they mainly concentrate on particular study areas in the discipline. As a result, it's still worthwhile to review recent progress in variable selection and explore some potential future difficulties.

Variable selection defines the process of choosing a variant of variants from a larger collection of characteristics that performs the best in terms of classification accuracy. The potential of variable selection to improve learning performance is essential in machine learning and pattern recognition. This implies that limiting the number of variants used for learning, can lower the cost of learning and improve learning accuracy when compared to utilizing the entire variant set. For many years, various variable selection methods have been suggested and debated. Finding the best variable selection technique in a specific data type or learning approach, on the other hand, remains a fundamental but challenging issue. To discover a solution to this issue, we compiled a list of recently suggested and debated variable selection algorithms and then compared each of them in this article via a series of tests [4].

7.2   Irrelevant Variants (Without Variant Selection): A Problem

A fundamental issue in machine learning is the identification of relevant character-istics and the removal of irrelevant ones. Before induction algorithm can generate predictions about new test cases outside the training data, it must first select which characteristics to utilize and which to reject. Intuitively, only the learner should use those characteristics that are "relevant" to the goal idea. Since researchers pointed out in their study of the subject, there have been a few efforts to define "relevance" in the perspective of machine learning. We will not take a stance on this problem since we will examine a range of methods. Instead, we'll concentrate on the job of identifying and choosing important characteristics (however defined) used in learning and forecasting. Many induction techniques, particularly those that work on logical representations, try to deal directly with the issue of attribute selection [5]. The methods for generating logical conjunctions, for example, essentially do nothing more than add or delete elements first from the concept description. The fundamental actions of more complex techniques for generating decision lists and decision trees are adding and deleting single characteristics. Weights are used to give degrees of significance to characteristics in certain non-logical induction techniques, such as neural networks and Bayesian classifiers. Furthermore, certain learning methods, such as the basic closest neighbor approach, disregard the problem of relevance completely. Induction techniques that extend well with domains with many irrelevant variants would be ideal. We'd want the collection complexity (the amount of training data instances required to achieve a certain level of accuracy) to increase steadily as the amount of irrelevant characteristics increases. Theoretical findings for limited hypothesis space search algorithms are promising.

The average-case analysis for Wholist, a basic conjunctive method, by Pazzani and Sarrett (1992) and Langley and Iba's (1993) study of the Naïve Bayes classifier by Langley and Iba both indicate that sample complexity increases at most directly proportional to the number of irrelevant characteristics. Theoretical findings for induction techniques that explore a broader set of idea descriptions, on the other hand, are less encouraging. Langley and Iba's (1993) average-case study of basic closest neighbor shows that, even for conjunctive target concepts, sample complexity increases exponentially with the number of irrelevant characteristics. Experimental investigations of closest neighbors support this conclusion, and additional experiments indicate that comparable findings remain true even for induction methods that explicitly choose variants [6]. For conjunctive ideas, for example, sample complexity for decision tree techniques seems to increase linearly with the number of irrelevant, but exponentially for parity concepts, because the evaluation metric in the latter case cannot differentiate essential from irrelevance characteristics (Langley and Sage, in press). These kinds of findings have prompted machine learning researchers to search for more advanced techniques for picking relevant characteristics. We provide a basic framework for this job in the following section and then look at some recent instances of work on this essential issue.

7.3   Algorithms for Machine Learning Were Utilized in This Research

Different machine learning techniques, such as decision trees (DT), neural networks, SVM, KNNs, and others, have been created to conduct learning on the datasets. Researchers used a DT as a classifier and particle swarm optimization as a variant selection approach. The suggested method employs both single-objective and multi-objective PSO algorithms. From the KDD Cup 99 dataset, eighteen effective variants having a detection rate of 89.52% were chosen. Researchers utilize the IWD method for variant subset selection, a nature-inspired optimization technique, coupled with such a SVM as a predictor for variant assessment. The algorithm was able to minimise the input dataset's forty five variants to a minimal of ten. On the KDD Cup 99 dataset, the suggested method has a precision and accuracy of 99%. For image operator chain detection, the researchers designed a new forensic algorithm-based two-stream CNN. This framework is unique in its approach due to the collection of local noise remnant verification and manipulating artefact verification. It is specifically intended to forensically identify a chain consisting of two picture operators that can learn tracking characteristics immediately from picture data. Researchers provided blended IDS with hierarchical anomaly filtration [7, 8]. The ID3 decision tree is used to categorize data into appropriate classes. KKN is a technique to assign a class label to an unidentified data point depending on the classifiers of its k-closest neighbors. On the KDD Cup 99 and NSL-KDD datasets, the suggested method is tested. On the KDD Cup 99 dataset, the overall accuracy is 95.68%, with a detection rate of 96.68% and a false alarm rate of 7.49%. With the NSL-KDD dataset, the suggested method has an accuracy of 93.95%, a detection rate of 95.5% and a false alarm rate of 11.63%. According to the particular architecture of the datasets, each of these algorithms employs a distinct approach. A suitable method should be selected based on the dataset structure to achieve good performance. Three machine learning methods are utilized to compare the impact of variant selection in this research. Random forest, Naïve Bayes and C4.5 are three distinct types of learning techniques. These methods are widely used in the machine learning field and have shown to be effective in practice.

7.4   Methodology for Feature Selection

The suggested technique, ACO with symmetric uncertainty (ACO-SU), is discussed in this section. ACO is ideally suited for this job since variant subset selection is a combinatorial optimization issue. The suggested approach uses ACO as a population-dependent variant subset selection approach, in which chosen subsets are assessed using an information-theoretic metric. Wrapper-dependent approaches utilize a learning technique to give a fitness measure of the chosen variants, while filter-dependent techniques depend primarily on selecting variants and selective sampling techniques in choosing a resultant variant subset. The use of meta-heuristics, such as ACO, in filter-dependent variant selection for an adaptive choice of a variant subset is investigated in this research [9].

A. *Uncertainty in a Symmetric Form*

The suggested approach utilized an information-theoretic measurement termed symmetric uncertainty (SU) to assess the value of built solutions. SU has various supremacy including the fact that it is symmetric in essence; thus SU(x, y) is the same as SU(y, x) lowering the amount of analogy needed when x and y are two characteristics. Unlike information gain, SU is unaffected by multivalued characteristics, and its values are normalized. The symmetric uncertainty equation is as follows:

1. Ant colony improvement

Researchers suggested ant colony optimization in their key paper. Ants are basic agents that iteratively build a potential solution. An ant represents a complete solution. Depending on the pheromone level on the route and the reliability of the answer, each ant builds a probabilistic solution. Every solution is assessed after a generation, and the travelled routes are improved. In successive generations, outcomes get more optimised.

B. *The Major Concerns for Using ACO to a Graph Representation Are Listed Below*

Each node must represent a variant. The network is configured using mesh topology because the characteristics gathered in a set of variants have no ordering importance. Each node is completely linked to every other node in the graph. The graph's traversal comes to a conclusion at the terminal node. Because each ant symbolizes a unique solution, it is possible to stop before completing the full tour of all nodes. Pheromone intensity and heuristic desirability are two variables connected with each connection of the search space for the variant subset selection issue [10].

2. Heuristic desirability and the feedback process

The previous generation of ants' experience is roughly recorded on the graphs in the kind of pheromone concentration. Positive feedback refers to the act of transmitting the experience to future generations of ants. The attractiveness of the nodes is another key factor for the ants. This data is also saved on the node-to-node connections. As a desirability heuristic, we utilized symmetric uncertainty. Following the construction of the candidate solutions, they are evaluated against one fitness function. According to the effectiveness of their different solutions, all ants are obliged to strengthen their route for future generations.

3. Constructing solutions and satisfying constraints

A variety of solutions are built-in each generation. The number of methods defines the number of generations of ants. This number is usually set for all generations. Each ant walks around the graph, recording all of the nodes it passes through in its local memory. An ant's gathered characteristics are assessed after it has completed its journey. Constraint satisfaction is the process of determining the viability of built solutions [11].

The initial action after loading a dataset is to calculate the symmetric uncertainty estimate of each attribute. After that, an ACO compatible search space is created, complete with parameter initialization. Each ant creates a solution, which is then assessed using a fitness function. Because it is possible to arrive at an early solution, all ants must update the pheromone trail of their routes according to the efficiency of their solutions [12]. If the ideal option of a generation remains unchanged after a number of generations, the loop ends, and the best solution acquired so far is generated as the permanent conclusion. We must, in general:

- Invent more clever methods for choosing an initial collection of characteristics to search from.
- Think of search-control techniques that take use of variant set space structure.
- Improvise frameworks for assessing the utility of different variant sets (better than the wrapper approach).
- Improve the efficiency of stopping criteria by designing better ones without jeopardizing the usefulness of variant sets.

The characteristics are ranked in decreasing order using statistical metrics. The top-performing variants must then be chosen using a threshold value. Because not all datasets have the same size, this isn't a very useful criterion for selecting variants, which may vary from one dataset to another. Even if 10%, 20%, 25%, or 50% of the best performing variants are chosen, the result will not be an optimum variant subset since top variants may only account for 5% of the having additional [13].

The problem with the aforementioned approach is that the number of characteristics to be chosen cannot be predetermined, as with a threshold value, or estimated, as with a percentage. For each dataset, the suggested method produces an optimum variant subset. As a result, the dataset's nature determines how many characteristics should be included in a final subset. Furthermore, the suggested approach favours subsets with few characteristics.

7.5　Approaches for Reducing Dimension

Due to high the computation complexity and memory consumption of high-dimensional data, classification algorithms struggle with it. Variant extraction (also called dimensional reduction directly or detailed design) and variant selection are two-dimensional reduction methods. The benefit of dimensional reduction is that no knowledge about the significance of a lone variant is lost. However, suppose only a small set of variants is needed and the primary variants are extremely manifold, information may be lost because few variants should be deleted during the variant subset choice procedure. Via contrast, the size of the variant space may frequently be reduced in variant extraction sans missing most of the information from the original space. The data type and application domain influence the decision between variant extraction and variant selection techniques [14].

7.5.1   Variant Picking

High-dimensional data contains characteristics that may be useless, deceptive, or duplicated, resulting in a larger search area, making it more difficult to analyze data and obstructing the learning process. The process of choosing the best characteristics from among the variants that may be used to distinguish classes is known as variant subset selection. The variant selection algorithm is a mathematical model which is triggered by a set of relevant criteria [15, 16]. Researchers provided an empirical survey of different variant selection approaches. Variant selection is often termed as a searching issue depending on a few evaluation benchmarks. The searching structure of feature selection approaches can be categorised into three groups: (i) random, sequential and exponential. (ii) Production of successors (subset): to produce successors, five alternative operators may be considered – compound, random, forward, backward and weighted. (iii) Evaluation methodology: successor evaluation may be measured using consistency evaluation, information or uncertainty, divergence, probability of error, dependence, and interclass distance.

7.5.2   Random Forest (RF)

Random forest denotes a set of tree predictors in which every tree is based on the values of a random vector that is sampled separately, and most of the trees in the forest have a similar distribution. The generalization error conforms to a limit as the forest tree gets bigger and larger. The generalization error of the forest of tree classifiers is determined by the relative tree intensities in a forest, and the correlation between them. The random forest perform a mix of inputs or randomly chosen inputs at each node to build the tree and improve accuracy [17]. This method helps reduce the correlation while also increasing the efficiency of the woodlands. The random forest with random variants is built at each node by dividing the limited number of independent variables and selecting variants at random. The technique is similar to CART, and the GINI index is used to determine which branch should be produced at each node. Random forest use bagging in conjunction with random variant selection. The tree is built on the freshly acquired training dataset using random variant selection, starting with the old training set and replacing it with the new training set. The trees in random forest are not trimmed but instead growing [18].

The use of bagging has mainly two benefits. To begin with, anytime random characteristics are used, the accuracy improves. Second, bagging produces generalization error estimates based on a mixed ensemble of trees, and correlation and

strength estimates. This estimate is done without the use of a bag. Out-of-bag estimations are based on merging of one-third of various classifiers from current combination. The mistake rate reduces as the number of combinations rises. As a result,
out-of-bag estimations will overstate the current error rate. As a result, it's critical to
go beyond the region where test error convergence occurs [19]. Out-of-bag estimates are likely to be unbiased, while cross-validation estimates have bias, although
the amount of the bias is unclear. The random forest algorithm utilizes two-thirds of
the training data to build the tree and one-third of the training dataset to assess the
error. Out-of-bag data refers to the third of the training dataset. Because RF does not
prune the tree, it is faster and more performance-focused than other decision tree
methods. Despite the fact that random forest functions better than the other tree
structures, its efficiency is more directed than some other decision tree methods
[20, 21].

7.5.3   Naïve Bayes

These classifiers are probabilistic classifiers with a powerful belief of naive independence as one of its characteristics. They are linked to the Bayes theorem. In
Naïve Bayes, knowledge is represented as probabilistic summaries. The Bayes theorem may be expressed mathematically as follows:

$$P(X/Y) = \frac{P(Y/X)P(X)}{P(Y)}$$

P(X) and P(Y) are two different types of occurrences.

The prior probability of X and Y is P(X) and P(Y). P(X/Y) is the posterior probability of witnessing an event X if Y is true. Given that X is true, the probability of
witnessing an event Y is known as P(Y/X). Researchers implemented the variant of
the Naïve Bayes that was utilized in this research. The nominal variant probabilities
are calculated from data, whereas the nominal variant probabilities are supposed to
come from a Gaussian distribution. A Bayes rule, also known as a Bayes classifier,
is a rule that uses the entire distribution of data to forecast the most likely class for
a particular occurrence. The mapping of Naïve Bayes is easy to grasp, mainly if log
occurrences are employed. The log chances get an additional scoring function with natural expression ability. The Bayes rule's accuracy is the greatest possible accuracy. Researchers discovered that removing identical characteristics
enhances the performance of the Naïve Bayes algorithm [22].

According to the researchers, when there are modest relationships in the data,
Naïve Bayes works well. The NB classifier has the benefit of needing the minimum
amount of data training time.

The C4.5 decision tree is utilized in this investigation. The C4.5 is denoted by
J.48 in Weka. The C4.5 algorithm is the replacement of the ID3 technique. The C4.5

constructs and prunes decision trees by using a top-down method. C4.5 builds the tree by identifying the best lone variant set, which allows the best single variant test to be performed at the tree's root node. The nodes of the tree are identical to the characteristics, and the branches are similar to the associated estimates.

The tree leaves correlate to the classes, and for ordering classification to a new instance, one must look at the variant checked at the tree nodes and then pursue the branch that corresponds to the estimates seen in the instance. The procedure ends when it reaches to the leaf, and the class is assigned to the instance [23, 24].

C4.5 builds decision trees using the greedy approach, which uses information theoretic measurements as a guide. The C4.5 divides the training examples into subsets which correspond to the number of attributes for choosing a characteristic for the tree's root. If the entropy of the class labels in the subset is lower than that of the entropy of the class labels in the entire training set, the attribute is divided to acquire information. C4.5 uses the gain ratio criteria to choose the property at the tree's root. The gain ratio criteria selects chooses the characteristics with best or average gain among their different qualities and those that maximize the gain ratio split by its entropy. In C4.5, subtrees are created by iteratively executing the technique, and the process is stopped after discovering that the supplied subset has just one class. The main dissimilarity between ID3 and C4.5 is that C4.5 prunes its DTs, resulting in simpler DTs and a lower risk of training data overfitting. C4.5 prunes a re-substitution mistake by exercising the upper limit of the confidence interval. When the leaf's estimation error is within one standard error of the node's estimated error, the node is replaced by the best leaf. C4.5 is used as a benchmark method when the effectiveness of machine learning techniques is assessed. As summarised by the method C4.5, anytime information is induced, it is accurate, resilient and quick. It also shows the benefit of performing extremely very well unnecessary and applicable characteristics, resulting in a significant improvement in accuracy when used with variant selection [25].

7.6 Supervised Filter Model Based on Relevance and Redundancy

This kind of approach is based on relevance (variant-class) and redundancy (variant-variant) scrutiny. These approaches utilize information measures, Pearson correlation, and Euclidean distance for relevance and redundancy analysis. An original collection of variants may be classified into four types: totally unrelated and noisy features, feebly admissible and unnecessary attributes, feebly specific and non-specific variants and highly appropriate variants [26]. Following the rough set theory, the highly relevant variant is also known as a mandatory attribute, and it is the core of the conditional feature set. The study of relevance and redundancy may be divided into two drawbacks: maximum (relevance $(Z;C)$) and minimum (redundancy (R')).

The classical clustering method is a technique for arranging objects in a logical manner. The symmetric uncertainty is used as a metric in this approach, which treats each minimal spanning tree as a cluster. FAST can handle high-dimensional data, but the quantity of chosen variants must not be set manually. MFC, a clustering variable selection technique depending on the minimal spanning tree, was also proposed by Liu. Unlike FAST, the information distance measure, or variance of information, is used by MFC. Researchers proposed the mRR variable selection technique based on hierarchical clustering. The conditional mutual information is used as the divergence metric - the higher the variant-variant variety, the lower the variant-variant redundancy. The redundancy metric, on the other hand, is often unable to be utilized as a direct indicator of diversity. The diversity measures in the techniques mentioned above are SU, conditional mutual information, and variation of information but not mutual information that is indicated in the MRMR criterion. Because unimportant characteristics are often grouped, the clustering-based approach may pick an irrelevant variant [27]. As a result, removing non-essential variants before variant clustering is preferable. Zou recently introduced the MRMD (Max-Relevance-Max-Distance) variant ranking technique, which ranks variants using a hybrid measure. Compared to the mRMR technique, this approach not only offers greater stability for variable selection, but it may have a lower running time.

7.7   Wrapper Model Under Supervision

The classification error or accuracy rate is used as the variant assessment benchmark in wrapper models. As the learning technique is incorporated in variable selection, the result of variable selection is often generated simultaneously with the result of the learning model. Compared to the filter model, the wrapper model may achieve better classification accuracy and tend to provide a lower variant magnitude; nevertheless, it has weak generalization capacity and higher time complexity. Researchers developed the SVM-RFE approach and used the wrapper method which utilizes SVMs to evaluate the efficiency of variants and builds a classifier having good performance criteria. Researchers developed the dual wrapper variable selection approach based on relationships [28]. This approach presents a rank criterion system based on class densities for binary data. A two-stage approach is utilized in the literature that uses a lower-expenditure approach based on rank variants and an expensive wrapper approach to remove unnecessary variables. Wrapper models often use the genetic algorithm and PSO, two random search methods. Researchers utilizes a decision tree to pick variants and a GA to discover a set of variant variants that reduce decision tree misclassification. The Fisher discriminant analysis and the GA were combined by various researchers. In chemical fault identification, the essential factors provide excellent results. Researchers evaluated the KNN classification function accuracy, and the variant weights of MPEG-7 images were ideally generated using a real coded chromosomal GA. The KNN algorithm is utilized as the evaluator, and an enhanced binary PSO is being utilized to perform variable selection for gene selection. Researchers presented a particle swarm optimisation variable selection approach with a convenient novel approached with classification accuracy including the quantity of chosen variants into account [29, 30]. The filter model is sometimes coupled to the wrapper model to create the hybrid variable selection technique. There are two phases to the hybrid variable selection: the first is utilizing the filter to decrease the variant space size by eliminating unrelated along with noisy variants, and the next is to use the wrapper to identify the optimum variant from the remaining variants. To create a filter-wrapper hybrid approach, various researchers coupled the mRMR method with a GA. Researchers developed FRF-fs, a hybrid filter-wrapper technique based on fuzzy random forest.

We provide a descriptive study to help you intuitively understand the variable selection technique. mRMR, ReliefF, SVM-RFE, information gain, and JMI are enforced to two high-dimensional gene expression datasets (Colon and CNS) that are frequently utilized in evaluating the achievements of variable selection techniques. Both datasets are available for download at Kaggle. CNS has 7129 variants, 58 specimens, and two classification labels, whereas Colon has 3020 variants, 72 specimens, and three classification labels. Because SVM-RFE is a wrapper model, the ml model is SVM. In addition, ten rounds of tenfold cross-validation are used in improving classification precision. The number of chosen variants is shown on the x-axis, while the average classification accuracy of every variable selection technique is depicted on the y-axis. We may draw the following three conclusions [31].

(a) The classification accuracy of such superior variable selection techniques is higher than 90%, demonstrating that variable selection approaches are successfully minimize data processing size.

(b) As the number of chosen characteristics grows, the classification accuracy improves. When the quantity of chosen characteristics tends to a specific threshold, the classification accuracy begins to stabilize. In most cases, the optimum number of chosen characteristics is low.

(c) SVM-RFE achieves a maximum accuracy of 99% that really is superior to other filter techniques. In general, the wrapper approach achieves higher classification accuracy. Information gain shows a time complexity of $O(n\,m)$, whereas ReliefF has a time complexity of $O(nm^2)$, where n is the variant quantity, and m is the dataset specimen size. JMI and mRMR have time complexity of $O(n^2\,m)$, while SVM-RFE has time complexity of O (maximum $(n,\,m)n^2)$.

7.8   Future Research on Feature Selection

Variable selection is a common optimisation issue, and the best answer may exclusively be found by conducting a search based on a set of criteria. For high-dimensional issues, researchers continue to use the heuristic approach with polynomial time complexity [32]. This chapter examined several typical supervised, unsupervised, and semi-supervised variable selection techniques, and their current applications in machine learning. Even though significant progress has been made in this field, there are still certain obstacles to overcome.

7.8.1   Machine Learning with Extreme Data

Smaller or larger specimens with higher dimensional and unbalanced class labels show extreme data. In many data mining areas, like text mining and information retrieval, big data having huge specimens and high dimensions has emerged. Weinberger, for example, looked at a collective spam email filtering job with 18 trillion distinct characteristics. The dimension may be very large in gene expression scrutiny, but the specimen magnitude is quite smaller (about 40). Various scholars concentrate their efforts on unbalanced categorization and associated issues. For extreme datasets, Novel variable selection techniques are required for large datasets, since conventional methods are often invalidated. Furthermore, for various kinds of extreme datasets, the study focus should be varied. For example, when dealing with high-dimensional datasets or big specimens, the time complexity of an algorithm is almost as essential as its accuracy. Meanwhile, while dealing with unbalanced datasets, variable selection techniques should concentrate on the accuracy of their minority class identification skills. Furthermore, while developing a variable selection technique, we must consider the accuracy, scalability and stability of the approach. The sensitivity of the computing capacity of the variable selection technique to the dataset is characterized as scalability. Accuracy, time complexity, and space complexity are all factors in computing performance. The scalability of the variable selection technique should be able to adapt to datasets of different magnitudes and with high order time complexity. The sensitivity of the variable selection outcomes to changes in the training set is described the stability [33]. The variable selection method's stability is deemed to be poor if the variable selection outcomes differ while connecting or removing certain specimens.

7.8.2   Variable Selection via the Internet

Contemporary variable selection relies on a static variant space, in which the variable selection input does not vary. However, data changes with time in many areas, like video data streams and network data streams. As a consequence, the acquired variant is continuously changing. Online variable selection refers to variable selection within this dynamic variant space. The choice of online variants is made independently of the online learning paradigm. It may be utilized for dynamic and unknown variant spaces; however, only specific variant spaces are suitable to the online learning paradigm. SAOLA, alpha-investing, FAST-OSFS, grafting and

OSFS are some of the most advanced online variable selection techniques. SAOLA (Towards Scalable and Accurate On Line Approach) accomplishes redundancy analysis via variant-variant correlation that substitutes variant searching in OSFS, and FAST-OSFS to eliminate repeated, significantly increasing time efficiency. LOFS, an open-source library for online variable selection, is openly accessible at Kaggle [34]. Three aspects of the efficiency of online variable selection techniques may be enhanced: method stability, repetition analysis, and time efficiency. The real-time analysis framework, in particular, places a premium on time efficiency. Fast online variable selection may be achieved with good algorithm design. Meanwhile, high-performance hardware may significantly increase processing speed.

7.8.3   Deep Learning (DL) and Variable Selection

Deep learning has distributed depictions of data by incorporating low-level characteristics in creating better higher-level variants. It is extensively utilised in voice recognition, picture processing and recognition, information comprehension and game intelligence. It is a significant breakthroughs in machine learning currently. Many models are utilized in deep learning, including DNN, CNN, RNN, and others. The following is a summary of the relationship between variable selection and deep learning. During the neural network training process, the irrelevant variant may consume many resources. To reduce the training time for deep learning techniques, variants should be carefully chosen. Individual training error reduction ranking is employed in picking nodes in every hidden layer that reduces of the deep neural network framework. Reconstruction fallacy is utilised to choose the input variables for a deep neural network, similar to a networking pruning method. Moreover, selecting variants at the incoming level aids in comprehending the structure of a complicated framework or trained replica [35]. The weight of an unimportant variant will be close to 0 while training a neural network. As a result, the deep learning technique may be used to pick variants. In remote sensing sequence classification/ recognition, a DL-based variable selection approach is presented that constructs the variable selection issue as a variant reconstruction issue. Deep learning with variable selection must be investigated additionally. For example, should we, for, e.g., contemplate deleting unnecessary and noisy variants from raw data and training the deep learning model with just relevant variants?

7.8.4   Selection Properties of Variant Variants

Generally, variable selection techniques must locate the best variant from a candidate set to characterize a learning system's goal concepts. The following factors must be addressed throughout the variable selection process. We must first identify the beginning point in variant space from the collection of complete variants, which affects the search direction. It's possible to start looking for variant variants with no variants. It may also begin with all (complete) characteristics. Forward selection is the first method, while backward elimination is the second [36]. Not only may we begin our search from the two places mentioned above. A set with half as many variants as the complete set or a set with a random range of variants may also be useful.

7.8.5   *Searching Engine Optimisation*

Optimal variant of characteristics may theoretically be discovered by assessing all potential variants, and this process is called exhaustive search technique. However, since an exhaustive search of the variant space requires searching entirely $2n$ potential variants of n variants, it is almost always unfeasible when dealing with huge numbers of variants. As a result, we must take a better pragmatic technique. Various simpler search methods have been devised; however, they do not ensure that the optimum variant of characteristics will be found. The computing cost of these search methods varies, as does the optimality of the variants they uncover [37, 38].

7.9   Classification Accuracy

We initially used the learning methods tenfold CV and LOOCV to evaluate the classification accuracy of datasets with complete variants. Then, we rearranged the datasets using chosen (reduced) characteristics for each technique, and assessed each method in the same way. We determined the difference in accuracies between the restructured datasets and the full-variant datasets by measuring the accuracy of those reorganized datasets. Most of the techniques performed better with a smaller variant set than with the complete one. For most of the datasets with two learning algorithms, the mRMR approach performed even better than previous methods. INTERACT also performed as well as mRMR when using the NB learning algorithm. For certain discrete data, however, the I-RELIEF technique performs poorly because of the method's variants [39, 40]. Because relief techniques rely on the distance between neighboring specimens, the weight of each variant may have been skewed in discrete instances. GA had a terrible track record as well. In contrast to the I-RELIEF results, GA performs poorly in microarray datasets. We created binary datasets from the three UCI datasets, arrhythmia, sonar, and ionosphere, to compare the performance of the CMIM technique with that of other methods. With those datasets, we repeated the tests [41, 42]. All approaches, as anticipated, decreased the number of characteristics. However, learning performance with chosen characteristics did not increase in all approaches. Other techniques did not enhance learning performance, whereas CMIM and mRMR did. The I-RELIEF technique, in particular, produced considerably poorer learning performance than the other methods. This may be related to the bias issue discussed with discrete datasets in the prior finding [43, 44].

7.10 Conclusion

We evaluated and tested eight alternative variable selection techniques, all of which were recently suggested and tested using publicly available data. Most of the techniques increased their outcomes related to the variant magnitude and classification precision. Compared to other techniques, we found that the mRMR method's outcome showed the most robust and consistent performance. Meanwhile, other techniques, such as I-RELIEF, were insecure, relying on the data or the learning strategy for stability. We want to fine-tune the parameters of our experiment in the future and compare the execution time of every technique to make a fair comparison. Each method must be correlated with its outcomes produced using the optimal parameter setting and its real running time. Finally, we want to conduct a new variant selection experiment.

References

1. Aristidis Likasa, Nikos Vlassis, Jakob J. Verbeek,"The global k-means clustering algorithm", the journal of the pattern recognition society, Pattern Recognition 36 (2003) 451–461, 2002.
2. Carlos Ordonez, "Clustering Binary Data Streams with K-means", San Diego, CA, USA. Copyright 2003, ACM 1- 58113-763-x, DMKD'03, June 13, 2003.
3. K. Wang et al., "A Trusted Consensus Scheme for Collaborative Learning in the Edge AI Computing Domain," in IEEE Network, vol. 35, no. 1, pp. 204–210, January/February 2021, doi:https://doi.org/10.1109/MNET.011.2000249.
4. Guha, D. Samanta, A. Banerjee and D. Agarwal, "A Deep Learning Model for Information Loss Prevention From Multi-Page Digital Documents," in IEEE Access, vol. 9, pp. 80451–80465, 2021, doi:https://doi.org/10.1109/ACCESS.2021.3084841.
5. Rohan Kumar, Rajat Kumar, Pinki Kumar, Vishal Kumar, Sanjay Chakraborty, Prediction of Protein-Protein interaction as Carcinogenic using Deep Learning Techniques, 2nd International Conference on Intelligent Computing, Information and Control Systems (ICICCS), Springer, pp. 461–475, 2021.
6. Guha, A., Samanta, D. Hybrid Approach to Document Anomaly Detection: An Application to Facilitate RPA in Title Insurance. Int. J. Autom. Comput. 18, 55–72 (2021). doi:https://doi.org/10.1007/s11633-020-1247-y
7. Lopamudra Dey, Sanjay Chakraborty, Anirban Mukhopadhyay. Machine Learning Techniques for Sequence-based Prediction of Viral-Host Interactions between SARS-CoV-2 and Human Proteins. Biomedical Journal, Elsevier, 2020.
8. Khamparia, A, Singh, PK, Rani, P, Samanta, D, Khanna, A, Bhushan, B. An internet of health things-driven deep learning framework for detection and classification of skin cancer using transfer learning. Trans Emerging Tel Tech. 2020;e3963. doi:https://doi.org/10.1002/ett.3963
9. Jiawei Han and Micheline Kamber, Data Mining concepts and techniques, Morgan Kaufmann (publisher) from chapter-7 'cluster analysis', ISBN:978-1-55860-901-3, 2006.

10. Dunham, M.H., Data Mining: Introductory And Advanced Topics, New Jersey: Prentice Hall, ISBN-13: 9780130888921. 2003.
11. H. Witten, Data mining: practical machine learning tools and techniques with Java implementations San-Francisco, California: Morgan Kaufmann, ISBN: 978-0-12-374856-0 2000.
12. Kantardzic, M. Data Mining: concepts, models, method, and algorithms, New Jersey: IEEE press, ISBN: 978-0-471-22852-3, 2003.
13. Michael K. Ng, Mark Junjie Li, Joshua Zhexue Huang, and Zengyou He, "On the Impact of Dissimilarity Measure in k-Modes Clustering Algorithm", IEEE transaction on pattern analysis and machine intelligence, vol. 29, No. 3, March 2007.
14. Nareshkumar Nagwani and Ashok Bhansali, "An Object Oriented Email Clustering Model Using Weighted Similarities between Emails Attributes", International Journal of Research and Reviews in Computer science (IJRRCS), Vol. 1, No. 2, June 2010.
15. Oyelade, O. J, Oladipupo, O. O, Obagbuwa, I. C, "Application of k-means Clustering algorithm for prediction of Students' Academic Performance", (IJCSIS) International Journal of Computer Science and Information security, Vol. 7, No. 1, 2010.
16. S. Jiang, X. Song, "A clustering based method for unsupervised intrusion detections". Pattern Recognition Letters, PP. 802–810, 2006.
17. Steven Young, Itemer Arel, Thomas P. Karnowski, Derek Rose, University of Tennesee, "A Fast and Stable incremental clustering Algorithm", TN 37996, 7th International 2010.
18. Taoying Li and Yan Chen, "Fuzzy K-means Incremental Clustering Based on K-Center and Vector Quantization", Journal of computers, vol. 5, No. 11, November 2010.
19. Tapas Kanungo, David M. Mount, "An Efficient k-Means Clustering Algorithm: Analysis and implementation", IEEE transaction vol. 24 No. 7, July 2002.
20. Zuriana Abu Bakar, Mustafa Mat Deris and Arifah Che Alhadi, "Performance analysis of partitional and incremental clustering", SNATI, ISBN-979-756-061—6, 2005.
21. Xiaoke Su, Yang Lan, Renxia Wan, and Yuming, "A Fast Incremental Clustering Algorithm", international Symposium on Information Processing (ISIP'09), Huangshan, P.R. China, August-21-23, pp: 175–178, 2009.
22. Kehar Singh, Dimple Malik and Naveen Sharma, "Evolving limitations in K-means algorithm in data Mining and their removal", IJCEM International Journal of Computational Engineering & Management, Vol. 12, April 2011.
23. Anil Kumar Tiwari, Lokesh Kumar Sharma, G. Rama Krishna, "Entropy Weighting Genetic k-Means Algorithm for Subspace Clustering", International Journal of Computer Applications (0975– 8887), Volume 7– No. 7, October 2010.
24. K. Mumtaz, Dr. K. Duraiswamy, "An Analysis on Density Based Clustering of Multi Dimensional Spatial Data", Indian Journal of Computer Science and Engineering, Vol. 1 No 1, pp-8–12, ISSN: 0976-5166.
25. A.M. Sowjanya, M. Shashi, "Cluster Feature-Based Incremental Clustering Approach (CFICA) For Numerical Data", IJCSNS International Journal of Computer Science and Network Security, VOL. 10 No. 9, September 2010.
26. Martin Ester, Hans-Peter Kriegel, Jorg Sander, Michael Wimmer, Xiaowei Xu, "Incremental clustering for mining in a data ware housing", 24th VLDB Conference New York, USA, 1998.
27. Sauravjyoti Sarmah, Dhruba K. Bhattacharyya, "An Effective Technique for Clustering Incremental Gene Expression data", IJCSI International Journal of Computer Science Issues, Vol. 7, Issue 3, No 3, May 2010.
28. Debashis Das Chakladar and Sanjay Chakraborty, Multi-target way of cursor movement in brain computer interface using unsupervised learning, Biologically Inspired Cognitive Architectures (Cognitive Systems Research), Elsevier, 2018.
29. Althar, R.R., Samanta, D. The realist approach for evaluation of computational intelligence in software engineering. Innovations Syst Softw Eng 17, 17–27 (2021). doi:https://doi.org/10.1007/s11334-020-00383-2.
30. B. Naik, M. S. Obaidat, J. Nayak, D. Pelusi, P. Vijayakumar and S. H. Islam, "Intelligent Secure Ecosystem Based on Metaheuristic and Functional Link Neural Network for Edge of

Things," in IEEE Transactions on Industrial Informatics, vol. 16, no. 3, pp. 1947–1956, March 2020, doi:https://doi.org/10.1109/TII.2019.2920831.

31. Debashis Das Chakladar and Sanjay Chakraborty, EEG Based Emotion Classification using Correlation Based Subset Selection, Biologically Inspired Cognitive Architectures (Cognitive Systems Research), Elsevier, 2018.

32. D. Samanta et al., "Cipher Block Chaining Support Vector Machine for Secured Decentralized Cloud Enabled Intelligent IoT Architecture," in IEEE Access, vol. 9, pp. 98013–98025, 2021, doi:https://doi.org/10.1109/ACCESS.2021.3095297.

33. CHEN Ning, CHEN An, ZHOU Long-xiang, "An Incremental Grid Density-Based Clustering Algorithm", Journal of Software, Vol. 13, No. 1, 2002.

34. Bock, Frederic E., et al. "A Review of the Application of Machine Learning and Data Mining Approaches in Continuum Materials Mechanics." Frontiers in Materials, vol. 6, 2019, p. 110. Frontiers, doi:https://doi.org/10.3389/fmats.2019.00110.

35. Amador, Sandra, et al. "Chapter 6 - Data Mining and Machine Learning Techniques for Early Detection in Autism Spectrum Disorder." Neural Engineering Techniques for Autism Spectrum Disorder, edited by Ayman S. El-Baz and Jasjit S. Suri, Academic Press, 2021, pp. 77–125. ScienceDirect, doi:https://doi.org/10.1016/B978-0-12-822822-7.00006-5.

36. Dabhade, Pranav, et al. "Educational Data Mining for Predicting Students' Academic Performance Using Machine Learning Algorithms." Materials Today: Proceedings, June 2021. ScienceDirect, doi:https://doi.org/10.1016/j.matpr.2021.05.646.

37. Dogan, Alican, and Derya Birant. "Machine Learning and Data Mining in Manufacturing." Expert Systems with Applications, vol. 166, Mar. 2021, p. 114060. ScienceDirect, doi:https://doi.org/10.1016/j.eswa.2020.114060.

38. Emami Javanmard, Majid, et al. "Data Mining with 12 Machine Learning Algorithms for Predict Costs and Carbon Dioxide Emission in Integrated Energy-Water Optimization Model in Buildings." Energy Conversion and Management, vol. 238, June 2021, p. 114153. ScienceDirect, doi:https://doi.org/10.1016/j.enconman.2021.114153.

39. Jimenez-Carvelo, Ana M., and Luis Cuadros-Rodríguez. "Data Mining/Machine Learning Methods in Foodomics." Current Opinion in Food Science, vol. 37, Feb. 2021, pp. 76–82. ScienceDirect, doi:https://doi.org/10.1016/j.cofs.2020.09.008.

40. Lord, Dominique, et al. "Chapter 12 - Data Mining and Machine Learning Techniques." Highway Safety Analytics and Modeling, edited by Dominique Lord et al., Elsevier, 2021, pp. 399–428. ScienceDirect, doi:https://doi.org/10.1016/B978-0-12-816818-9.00016-0.

41. Ma, Ying, et al. "Meta-Analysis of Cellular Toxicity for Graphene via Data-Mining the Literature and Machine Learning." Science of The Total Environment, vol. 793, Nov. 2021, p. 148532. ScienceDirect, doi:https://doi.org/10.1016/j.scitotenv.2021.148532.

42. Yang, Xin-She. "Chapter 16 - Data Mining and Deep Learning." Nature-Inspired Optimization Algorithms (Second Edition), edited by Xin-She Yang, Academic Press, 2021, pp. 239–58. ScienceDirect, doi:https://doi.org/10.1016/B978-0-12-821986-7.00023-8.

43. Zhao, Qingkun, et al. "Machine Learning-Assisted Discovery of Strong and Conductive Cu Alloys: Data Mining from Discarded Experiments and Physical Features." Materials & Design, vol. 197, Jan. 2021, p. 109248. ScienceDirect, doi:https://doi.org/10.1016/j.matdes.2020.109248.

44. Zou, Chengxiong, et al. "Integrating Data Mining and Machine Learning to Discover High-Strength Ductile Titanium Alloys." Acta Materialia, vol. 202, Jan. 2021, pp. 211–21. ScienceDirect, doi:https://doi.org/10.1016/j.actamat.2020.10.056.

Chapter 8
Data Mining-Based Variant Subset Features

8.1 Introduction

In high-dimensional data analysis, visualization, and modeling, dimensional reduction is a widely used preprocessing technique. Variant selection is an easy method to decrease dimension; it chooses just that information relevant that has the essential information for addressing the issue. Variant extraction is a broad technique that consists of trying to create a translation of an input space onto a lower-dimensional subspace that retains most of the useful information [1]. The goal of variant extraction and selection techniques is to boost performance like predicted accuracy, visualisation and comprehension of acquired information. Variants may be classified as important, unimportant or inessential in general [2].

The correlation coefficient, INTERACT, SVM-REF, RELIEF, CMIM, BW-ratio, genetic algorithm, PCA, non-linear PCA, and correlation-based variant selection are only a few of the methods that have been developed. Given the large variety of current variant selection and extraction techniques, it is essential to have a benchmark to rely on when deciding which approach to employ in various circumstances [3, 4]. A study of these approaches is conducted depending on a literature review to determine the necessity of different variant selection and extraction methods in specific situations depending on assessments conducted by researchers to determine how these approaches help to enhance the classification algorithm predictive precision.

There are three types of variant selection techniques: filters, wrappers and engrained approaches. Because the variant selection method tailored is for the classifier to be employed, wrapper techniques outperform filter methods. Wrapper techniques, on the other hand, are too costly to employ for vast variant spaces due to their high computational cost, so each variant set must be evaluated with the trained classifier, slowing down the variant selection approach. When contrasted to wrapper approaches, filter techniques show a lower computational expenditure and are faster,

© The Author(s), under exclusive license to Springer Nature Switzerland AG 2022   177
S. Chakraborty et al., *Data Classification and Incremental Clustering in Data Mining and Machine Learning*, EAI/Springer Innovations in Communication and Computing, https://doi.org/10.1007/978-3-030-93088-2_8

but they provide lower classification reliability and are more suited to high-dimensional datasets. Hybrid approaches, which include the advantages of both filters and wrappers approaches, have been organized [5]. A hybrid method employs both an online assessment and a variant subset performance assessment algorithm. As illustrated, filter techniques may be divided into variant-weighting approaches and subset learning approaches. Variant-weighting algorithms give weights to specific characteristics and rank them according to their relevance to the goal idea. Relief is a very sound algorithm that evaluates relevance.

8.2   Review of the Literature

Techniques for reducing dimensional have become more important in the medical sector (automated application). In today's world, a tremendous quantity of data is produced in the medical field. It covers a patient's symptoms as well as numerous medical test results that may be produced. The terms "variant" and "input variables" are interchangeable. The characteristics of a medical diagnostic, for example, may include symptoms, which are a collection of factors that categorize a patient's health condition (e.g., diabetic retinopathy indication of dry, or wet age-based macular degeneration) [6, 7]. Here, a literature overview is shown for several commonly utilized variant selection and extraction techniques for ophthalmologists in the detection and diagnosis of numerous eye disorders (glaucoma, diabetic retinopathy, and particularly for automated diagnosis of age-based macular degeneration). The primary goal of the study is to raise awareness among doctors about the advantages and, moreover, the need to use dimensional reduction methods. It is necessary to be aware of the different benefits of dimensional reduction methods to profit from them for the aim to increase the accuracy of learning algorithms [8]. The following benefits of variant selection were given.

- It lowers the variant space to minimize storage needs and speed up the algorithm.
- It eliminates redundant, or noisy data.
- The immediate benefits for data analysis activities include a reduction in the time it takes for learning algorithms to execute.
- Improves the accuracy of the data.
- Improves model's accuracy.
- Variant set minimization to save time and resources during the next cycle of data gathering or during use.
- Increased the prediction accuracy via improved performance.
- Data visualisation or data comprehension to learn more about process that produced the data.

Researchers developed a method for extracting image-based characteristics from digitized retinal images to identify AMD. An ophthalmologist categorized 100 pictures into 12 groups based on the disease's visual variants. To extract characteristics, independent components analysis (ICA) was utilised as input to a classifier [9]. It was demonstrated that ICA may reliably identify and describe characteristics in funds pictures and extract quantitative variants out of each image implicitly to determine the phenotype.

The impacts of class noise misclassification on supervised learning for areas of medicine have been studied by various researchers. A study of relevant surveys on learning from noisy data was addressed, suggesting that variant extraction be used in preprocessing phase to reduce the impact of class noise on the learning experience. The filtering methods specifically handle noise. Many filtering methods have been outlined that researchers have found to be helpful [10]. The same researchers, on the other hand, have identified several practical problems with filtering methods. One issue is that without the assistance of an expert, distinguishing noise from exceptions (outliers) is difficult. Another issue is that a filtering method may require an anticipated amount of noise as an input variable, which is seldom known for specific datasets. Variant extraction methods (using PCA) are preferable for noise tolerance techniques since they assist in preventing implicit overfitting inside learning procedures [11]. The use of variant extraction methods prior to supervised learning allows for a reduction in the detrimental impact of mislabelled occurrences in the data. Diagnosis technique based on ANN is shown effective results.

The results of the experiments indicate that the combination of artificial neural network-singular value decomposition and principal component analysis is a viable diabetes diagnosis method with low computing cost and good accuracy. Because of the noisy data, variant extraction techniques were shown to be more appropriate for the automatic identification of ophthalmology illnesses than variant selection techniques. The majority of biomedical datasets include noisy data rather than useless or redundant data [12, 13].

8.3   Extraction of Variants

Variant extraction entails a few modifications of the primary variants to produce more important variants. "Variant extraction is usually understood to imply the creation of linear combinations Tx of continuous variants that have high discriminating power across classes", according to researchers. Finding an appropriate description of multivariate data is a major issue in neural networks study and other fields such as artificial intelligence. In this context, variant extraction may be used to minimize complexity and provide a straightforward depiction of data, with every parameter in variant space represented as a weighted sum of the primary input variable [14]. PCA is a popular and frequently used variant extraction method. PCA is suggested in various forms. PCA is a non-parametric technique for transforming the principal facts from a collection of unnecessary data. PCA is a data transformation approach that reduces redundancy (computed through the covariance) while maximizing knowledge (computed through the variance) [15, 16].

The impact of most of the dimensional reduction approaches (consisting of variant subset selection utilising information gain (IG) along with wrapper techniques and variant extraction of PCA techniques on classification performance) is empirically tested on two different categories of datasets (email data along with drug discovery data). The outcomes indicate that PCA variant extraction is significantly dependent on the kind of data. In an approach for selecting variants for both types of data, the wrapper technique has a more acceptable impact on the classifier than IG [17].

The outcomes of the experiments demonstrate the significance of a dimensional reduction procedure. Compared to variant extraction techniques, wrapper techniques for variant selection generate the lowest variant subsets with high competitive classification precision. Wrappers, on the other hand, are much more computationally costly than variant extraction techniques [18]. Researchers developed bi-level dimensional reduction techniques to improve classification performance with combined variant selection and variant extraction techniques. They suggested two techniques: in the first level of dimensional reduction, variants are chosen depending on mutual correlation, and in the second level of dimensional reduction, variants were extracted depending on mutual correlation. Selected variants are utilized in the second level to extract variants using PCA and LPP. The proposed technique was tested on various common datasets to see how well it worked. The collected findings demonstrate that the suggested system outperforms single-level dimensional reduction methods [19, 20, 22].

8.4   Variant Selection and Extraction Approaches

Researchers depict various fundamental FSAs and their drawbacks. The Chi-squared test is usually the utilized statistical test to find the value a variant's event diverges from the predicted distribution if the variant's event is taken for granted irrespective of the class values [21]. The root of square discrepancies between coordinates of two objects is measured by the Euclidian distance. The benefit of the technique is that including various items in the survey that can be outliers has no consequence on the distance [23]. Dissimilarities in magnitude among the dimensions wherein the distance is calculated may find some consequence on Euclidian distance. The t-test evaluates that the two groups' mean is significantly different from others. The approach is favored anytime between two groups' averages [24]. Moreover, it is well relevant for the post-test only. Information gain is a measurement that differentiates the improvement in entropy, while a variant is available vs. during its absence. Correlation-based variant selection finds out variant subsets depending on how inessential the variants are. The objective of the evaluation is to identify subsets of features that are strongly connected with the class singly but consist of minimum intercorrelation. The priority of a set of characteristics expands when the correlation between variants along with class rises and decreases as the intercorrelation expands [25, 26]. It is often utilized in connection with search methods like genetic search, backward elimination, best-first search, bidirectional search, and forward selection to find the optimal variant subset. Pure relevance-dependent variant-weighting methods do not fulfill the requirement for variant selection, particularly better in variant selection for data having high dimensionality where there could be numerous duplicate variants [27].

8.5 Choosing of Variant Subset Features

Subset search techniques examine possible variant subsets depending on an evaluation measure that calculates the excellence quality of every subset. Utilizing different datasets from the website, the achievement of several variant selection techniques was evaluated. The number of reduced variants with their results on learning accomplishments was evaluated, scrutinized, and differentiated utilizing various commonly utilised techniques. The variant selection approach must pick the optimal variant subset from variant space to achieve the goal of the learning approach [28, 29]. In the variant selection process, various factors should be calculated:

1. The point of origin
2. The searching technique
3. The subset assessment and
4. The ending criteria

Researchers used different variant selection techniques. Mutual information (MI) of two random variables is utilized in mRMR (minimal redundancy and maximal relevance). MI is a parameter that shows the dependency of the two variables on each other. This approach grants the MI here between variant and a class that is the variant's connection to the class, along with the MI between variants to be the variant's redundancy. I-RELIEF is a popular weighting (ranking) technique to determine the significance of traits in samples that are closest to the target sample. Relief finds the closest sample in a similar category's variant space, dubbed the hit specimen, and finally measures the distance between the target and hit specimens [30]. Moreover, it traces the nearest specimen of the reverse category, called a miss specimen, and executes the same job. The distance between those calculated distances calculates the weight of the achieved variant. This important technique was split into various types [31].

The I-RELIEF approach decreases the bias of the existing RELIEF approach. Utilizing the conditional mutual information, Conditional Mutual Information Maximization (CMIM) finds a variant subset with the maximum relevancy to the target class. The two extracted variants and the output classes should be binary in CMIM [32].

The correlation coefficient approach measures how fruitfully a certain feature promotes class separation. For every specimen of the two classes, ranking criteria are used to rank every feature depending on their mean and the standard deviation. The between-within ratio (BW-ratio) calculates the ratio of between-group to within-group sums of squares for every characteristic and chooses the one with the highest BW-ratio. Based on its connection to the class, a single variable may be deemed unimportant, but when combined with different characteristics in the variant space, it results in being an extremely effective approach. INTERACT techniques take variant interaction into account. Backward elimination with consistency contribution measurement is used to identify interacting variants in this method. The C-consistency of a variant measures how much the removal of that variant will impact consistency; for example, the C-consistency of an unimportant variant will be 0.

A randomized method is used in the genetic algorithm (GA). A GA refers to a subset of evolutionary algorithms that utilized evolutionary biology methods, including inheritance, mutation, selection, and crossover [33]. A binary string shows every variant set for variant selection issues. Backward elimination is achieved utilizing the recursive variant elimination (SVM-RFE) wrapper approach. SVM-REF uses the weight vector w as a ranking criteria to identify m variants which finds the greatest margin of class separation. PAM (classification analysis of microarray) is a statistical technique of class prediction using reduced centroid gene expression data [34, 35]. The closest shrunken centroid approach finds groups of genes that describe a class in a better manner. Researchers present an experimental study of the above-stated variant subset selection methods.

Seven datasets via UCI machine learning were collected and precompiled for discrete variant selection, including lung cancer, leukemia, and five additional datasets. Naïve Bayes and LIBSVM algorithms, were utilized to assess variant subset selection methods. Almost all of the methods described above worked well with a smaller variant subset. For most datasets, the mRMR technique outperformed all other methods. For discrete data, however, the I-RELIEF technique performed poorly. Microarray findings were similarly unsatisfactory because of GA. Compared to the techniques that individually manage variant duplication and/or irrelevant variants, extracted variants that handle the removal of both redundant and irrelevant variants simultaneously are considerably more robust and helpful for the learning process [36, 37].

Because of the noisy data, variant extraction techniques were also more appropriate for the automatic identification of ophthalmology illnesses than variant selection approaches. As the majority of biomedical datasets include noisy data rather than useless or redundant data, the variant selection is a tool that may be used to eliminate unnecessary and/or superfluous variants in a variety of applications. There is no one technique for selecting variants that can be used in all applications. Some techniques were utilized to remove unnecessary characteristics while avoiding duplicate ones [38, 39]. Pure relevance-based variants weighting algorithms do not effectively meet variant selection requirements. Subset searching approaches look for possible variant subsets depending on an evaluation metric that calculates how good every subset is. The monitoring key and the correlation measure are two current assessment methods that have been proven to be helpful in eliminating both irrelevant and duplicate characteristics. Experiments indicate that the number of iterations needed to identify the optimal variant subset is usually at least quadratic to the range of variants. As a result, current subset search methods with quadratic or greater time complexity in terms of dimension do not have sufficient scalability to cope with large-dimension data [40].

Filters and wrappers are two types of variant selection techniques. Since the variant selection procedure is tailored to the classification method, wrapper techniques usually outperform filter methods. However, if the selection of attributes is high, they are generally much too costly to utilize since each variant set must be assessed with both the trained classifier. Filter techniques are considerably quicker than wrapper methods, making them better suited to large datasets. To cope with high-dimensional data, methods in a hybrid paradigm have recently been developed to incorporate the benefits of both models. And there aren't many ways for dealing with noisy data [41, 42]. Variant extraction techniques have been suggested as a preprocessing step to reduce the impact of class noise on the learning process. According to studies, the classification accuracy obtained with various variant reduction methods is mainly dependent on the data type. When opposed to techniques that discretely manage variant redundancy and/or irrelevant characteristics, variant selection approaches that control both superfluous and unimportant variants simultaneously are considerably much vigorous and helpful for the learning approach. Both approaches have the goal of reducing variant space for boosting data analysis. His aspect becomes even more essential when dealing with real-world datasets, which may include hundreds or thousands of characteristics. The most significant distinction between variant selection and extraction is that the former reduces dimension by choosing a subset of variants without transforming them. In contrast, the latter lessens dimension by calculating a transformation of the original variants to produce other, more significant variants. This chapter shows traditional techniques, their subsequent improvements and some intriguing applications for variant selection. Variant selection enhances understanding of how it works under examination by highlighting the most significant characteristics impacting the phenomena under investigation. Furthermore, the computation speed and accuracy of the chosen learning machine must be evaluated, since they are critical in ML and data mining approaches.

One of the most challenging issues in pattern recognition and ML is variant subset selection. Choosing only the characteristics that help forecast a target notion, such as class, is known as variant subset selection. Data collected from various sources is not vetted for any specific purpose, such as classification, clustering, or anomaly detection. When data is put into a learning algorithm, the outcome degrades. The suggested approach is a variant subset selection strategy based on pure filters that is low in computing cost and high in classification accuracy. Furthermore, in most instances, the suggested approach needs fewer characteristics in addition to excellent accuracy. The problem of variant ranking and threshold selection is addressed in the suggested approach. The suggested approach chooses the number of variants based on the value of each variant in the collection. Extensive testing is carried out using a variety of benchmark datasets and three well-known classification methods. Previously, data was manually converted into knowledge using analysis of data and interpretation methods [43].

This manual examination was highly subjective, time-consuming and expensive. Manual data analysis became laborious and impracticable in many areas when data production and recording increased dramatically. The necessity for an effective and efficient knowledge discovery method arose due to this. The world's information is expected to double every 20 months. These data explosions results from digital data collection, production, storage, and retrieval. Because data is produced at a quicker rate, massive quantities of data weren't examined due to a lack of effective analysis methods. Furthermore, since it is challenging to evaluate data in its totality, the term "knowledge discovery in databases" (KDD) was coined to describe the process of automatically or semi-automatically analyzing data. Data reduction is one of the most critical aspects of the KDD process is data reduction.

Datasets may include duplicate and unnecessary characteristics since data is not gathered with a particular goal, such as categorization. The presence of such traits may stymie the process of knowledge discovery. Furthermore, the time spent analyzing such characteristics may add to the total processing cost. One of the most common data reduction techniques is variant subset selection (FSS). This step's primary goal is to identify valuable characteristics and eliminate those that are either unnecessary or redundant.

An irrelevant characteristic does not offer any information that may be used to forecast the target idea, and a duplicated variant does not supply any extra details that may help to predict the law that aims. FSS assists in various ways, including reducing unnecessary characteristics to reduce computation time and data repository storage, improving predictive assessment, preventing overfitting and providing a much more accurate definition of the target idea. Variant selection is an optimizer in which a variant set comprising N variants is too big to explore exhaustively. Filter-based techniques and wrapper-dependent approaches are the two major types of selection-based algorithms [44].

Surrogate classifier measures, such as statistical metrics, are used in filter-based techniques to conduct FSS. Compared to other techniques, filter-based methods are lightweight and have low operating computational cost. In filter-based techniques, statistical measures that leverage the inter-variant connection between various variants are extensively used in filter-based techniques. Wrapper-based techniques, on the other hand, utilize a learning approach to assess the utility of a chosen variant subset, and certain search methods, such as community searching, to explore the possible variant space. Even though wrapper techniques are much more costly than filter approaches, wrapper approaches are thought to be a more accurate in terms of prediction. A variety of proposed variant selection methods have been suggested in the literature, with a few of them relying on computational intelligence being included below.

Variant ranking depending on a similarity metric is used in filter-based techniques. A variety of similarity measures have been suggested in the literature. Distance measurements, information theory measures, dependency indicators, and consistency type of measure are the four subtypes of these metrics.

Wrapper-based variant selection makes use of some learning algorithm's classification accuracy. A hybrid classification method was suggested by several researchers. It's a two-part procedure. In the first stage, the individual variants' symmetric uncertainty (SU) is computed, and those with SU less than the threshold value are eliminated. On the left-over variants, the second phase employs genetic algorithm-based searching. Tenfold cross-validation is executed to assess the quality of the variant subsets using the Naïve Bayes classifier and SU. Researchers suggested an ant colony optimization (ACO) and mutual information-driven variant selection for equipment failure detection. The variant subset is evaluated using a regression approximation method and MSE.

Researchers developed an ACO and mutual information-driven hybrid variant subset selection method for weather forecasting. Researchers also presented a fuzzy-rough data minimization method based on colony optimization and C4.5, in which ACO is utilized to identify a fuzzy-rough set. The halting criteria are defined as a fuzzy-rough-based dependence measure. Using particle swarm optimization, researchers presented a rough set-dependent variant selection. Standard PSO is developed into binary PSO since it is utilized for continuous optimization issues. For rule induction and subgroup optimisation, and the size of the reduction, LEM2 is utilized. Researchers developed a variant selection method dependent on consistency. During the training, examples are mapped onto the subset of attributes, determined using the degree of coherence in the class values. Because any subset's consistency is never lesser than the whole attribute set, it's common practice to combine this subset evaluator with a random searching technique to find the lowest subset having coherence the same as the entire set of attributes. The GA is utilized to generate variant subsets, and the previously stated consistency-based subset assessment.

8.6 Unsupervised Variable Selection

Unsupervised variable selection techniques attempt to balance the natural categorisation of data along with enhancing the clustering precision by selecting a variant depending on either clustering or evaluation criteria. Relying on if they use clustering approaches, unsupervised variable selection techniques may be unsupervised filter or wrapper variable selection techniques.

8.7 Model of an Unsupervised Filter

The unsupervised filter variable selection techniques pick variants as per the nature of the data variants. The variable selection method does not use clustering or learning techniques, which reduces the time of clustering and algorithm complexity. The unsupervised filter variable selection technique uses all of the statistical performance of the training data as the assessment metric, making it flexible and appropriate for big datasets. However, since the assessment conditions are free of the particular clustering method, the clustering performance of the chosen variant is typically worse compared to the wrapper model. Researchers suggested that entropy be used to assess the importance of variants, with the trace criteria being utilized to pick the best relevant variant. Various researchers proposed different unsupervised filter variable selection approach. It classifies the specimens and determines the number of clusters using a competitive learning method and thus splits the primary variant set into multiple variant variants. The mean dispersion in class and the mean scatter distance between classes are calculated using a judgment function. To identify the candidate variant, the utility of the judgment function for every variant is computed, and the variant that maximizes the judgment function estimate is chosen. Moreover, the correlation coefficient between the candidate variant and the chosen variant is computed. The candidate variant is eliminated if the correlation coefficient is higher than 0.80. Researchers used the probability density of various variant domains in an unsupervised setting to examine the distribution of variant data. The data distribution connection amidst the variants selects the variant. Researchers devised an unsupervised variable selection approach that measures variant similarity using the most significant information compression index. Since it does not need a search, this technique is fast and adaptable to datasets of various sizes. Various researchers developed an unsupervised attribute clustering technique and an unsupervised variable selection technique. It first constructs an attribute distance matrix by calculating the maximum information coefficient for every attribute pair and then groups all attributes utilizing the optimum K-modes clustering technique to identify K-modes attributes as variants of every cluster. Moreover, the quantity of clusters is immediately calculated. Researchers developed the FSFC approach, a clustering-dependent unsupervised variable selection technique that works in the same way as clustering-dependent supervised variable selection approaches. FSFC additionally performs better with datasets with a lot of dimensions. Unsupervised variable selection methods have also been proposed, together with filter algorithms depending on the Laplacian score. These methods are concerned with the data clusters' local topology. Researchers developed a technique depending on the notion that data belonging to a similar class ought to be near together. On the other hand, the Laplacian score is being utilized to assess the significance of the characteristics. Researchers used GA with Sammon's stress function to pick variants, maintaining the original data's topological structure in the smaller variant space.

8.8 Unsupervised Wrapper Approach

The validity of variable selection is adjusted using a clustering-based approach in the unsupervised wrapper variable selection technique. The ultimate optimum variant will be the variant with the most excellent clustering performance. The wrapper method's variant clustering performance is typically superior to the filter method's variant clustering performance. However, since the clustering algorithm must assess each variant, this approach has a high computational cost, which may be an issue when working with large-scale data. Furthermore, regardless of whether the variable selection is made in every cluster or only one, the unsupervised variable selection technique may be split into universal and logical wrapper models. Researchers used EM clustering to explore the wrapper framework by selecting variant variants. The maximum likelihood variables of a finite Gaussian function are estimated using the EM method. Then, scatter separability, and maximum likelihood are employed to assess potential variant subsets. The CLAS-SIT, a conceptual information hierarchical clustering method developed by researchers, now includes the variable selection. According to the clustering capacity of variants, this unsupervised variable selection technique looks for the best variant from the most important variants. The variant search will continue until the newer variant chosen cannot affect the clustering outcomes. The goal of the approach is to enhance variable selection's validity and forecast accuracy. To identify variant variants, researchers used sequential forward and backward searching techniques. The variant is evaluated using a clustering method, and the best variant is chosen based on clustering accuracy. In the document clustering problem, researchers devised an object evaluation function in selecting the variant and utilized the Bayesian statistical assessment methodology to determine the optimum quantity of clusters. The researchers developed a polynomial framework related to every cluster and used hierarchical clustering on it. For variable selection, researchers developed W-K-means, an enhanced K-means clustering method. Variant weighting is utilised to direct the average clustering approach so that it concentrates on essential characteristics in preference to relying on every variant equally. The W-K-means method is employed to cluster the datasets to generate a weight set for every variant. The variant set is then chosen based on the variant's weight, and the chosen variants are deleted from the dataset. Moreover, the datasets are clustered using W-K-means or different clustering methods to get the final clustering outcomes. Researchers proposed an unsupervised variable selection method for specimen-dependent gene expression data clustering. The suggested research utilizes PSO to find the optimal variant and employs the K-means method to assess the variants.

8.9 Selection of Semi-supervised Variants

Semi-supervised learning approach utilizes Su to enhance the learning capacity of the learning approach skilled by S given the dataset $S = [S_i, S_u]$, where S_i is the dataset having class labels and S_u denotes the dataset devoid of class labels. In semi-supervised learning, semi-supervised variable selection techniques, mostly filter approaches, enact a significant job. Several semi-supervised variable selection techniques use score functions, which are categorized into four types: Laplacian score, constraint score, variance score and Fisher score. For variable selection, semi-supervised variable selection techniques depending on Laplacian score integrate Laplacian criteria with outcomes in information. These are graph-related techniques for constructing the neighborhood graph and assessing the variants' capacity to preserve the local data structure. The characteristics of the Fisher criteria, and the local structure and distribution information of labeled and unlabelled data, are used in semi-supervised variable selection techniques depending on the Fisher score to choose the variants with the finest discriminant and locality-preserving abilities. Pairwise constraints along with the local characteristics of labeled and unlabelled data are used in semi-supervised variable selection techniques depending on pairwise constraints to assess the significance of variants based on their constraint and locality-preserving capacity. Individual characteristics are typically assessed using the techniques described above. Existing research emphasizes variant redundancy analysis while developing semi-supervised variable selection techniques. Researchers presented a filter dependent on a restricted Laplacian score, where the repetition is eliminated once the corresponding variants are chosen. Investigators proposed a semi-supervised filter variable selection technique depending on information theory termed SRFS, in which unlabeled data are used in the Markov blanket as labelled data by relevance gain. In the meantime, further study of semi-supervised variable selection techniques for regression issues is required.

8.10   Evaluation of Variants

We must assess variants of candidate characteristics once they have been generated. One technique, known as the filter method, involves employing a metric function to determine a variant's capacity to distinguish across data classes. Another option is the wrapper technique, which uses a learning approach as the metric function. It assesses each produced variant's quality (e.g., classification accuracy) by executing the learning approach to the data.

8.10.1   Specification for Stopping

Additionally, we should provide conditions for terminating the search. For example, when no other options increase the prediction of classification accuracy, one may cease adding or deleting variants or when the quantity of chosen variants exceeds a precalculated threshold. The best variant of the candidates we found throughout the search may subsequently be chosen.

8.10.2   Experimenting with Variable Selection Techniques

We selected eight techniques for selecting variant variants. We attempted to select the most updated techniques while maintaining further diversity in the processes' techniques as feasible. All of the techniques are open to the public.

mRMR

The mutual information of two random variables denotes a number which indicates how dependent the two variables are on each other. The mutual information between a variant and a class is used by the mRMR method to determine the variant's relevance for the class. The best way to find a variant set S_1 that meets the following criteria is as follows:

Maximum $X(S_1, c_1)$;
$X = 1/S_1 \sum y(z_i ; c_1)$; where $z_i \, \mathcal{E} \, S_1$

and $y(z_i ; c_1)$ denotes the mutual information between characteristic z_i and c_1. The mutual information between variants is also used by mRMR as variant redundancy.

8.10.3   I-RELIEF

Researchers originally developed RELIEF, a very well variant-weighting (ranking) method. The fundamental concept is to assess the significance of characteristics for areas throughout target specimens.

RELIEF identifies the closest specimen in variant space of a similar category, termed "hit" specimen, for each target specimen and then calculates the distance from the target to the hit specimens. Additionally, it locates the closest specimen from the other type, termed as the "miss" specimen, and repeats the process. The difference between the measured distances is used by RELIEF to calculate the weight of the target variant.

Several variations of this fundamental method have been developed. We used the iterative method to RELIEF (I-RELIEF) in our studies, which minimizes the bias of original RELIEF.

8.10.4   Interactive

The INTERACT method takes variant interaction into account: based on its connection with the class, a single characteristic may be deemed irrelevant, but when coupled with other qualities, it may become important. Backward elimination and assessment of consistency contribution (C-contribution) are used to identify interacting variants in INTERACT. The consistency contribution of a variant is a measure of how much the removal of the variant would impact uniformity. (In other words, an unimportant variant's C-contribution is 0.) INTERACT begins with the entire variant collection and removes variants one by one dependent on their C-contributions via backward elimination: if a variant's consistency contribution is less than the threshold (a predetermined tiny number), it is removed from the variant collection.

8.10.5   Genetic Algorithm (GA)

A well-known randomized method is a genetic algorithm (GA). It is a type of evolutionary methodology that employs evolutionary biology-inspired techniques, like inheritance, mutation, selection, and crossover. Binary string denotes each variant in variable selection issues. The presence of variant X_n in the variant set is indicated by bit 1 of the Nth bit. In a genetic algorithm, a fitness function is a kind of objective function that measures the optimality of a solution. Here in the experiment, we utilize the wrapper method, i.e., the accuracy of learning algorithms will be used to assess fitness function. To evaluate the calculation of each variable selection method, we utilized Weka. Weka is a popular Java-based ML program. We also used two learning algorithms to assess chosen variant variants: the Naïve Bayes (NB) and LIBSVM methods (LIBSVM). First, we look at how variable selection techniques decrease the numberof variants. The set of attributes in a dataset should be reduced since this reduces the complexity and learning time. All approaches effectively reduced the number of characteristics, as can be shown.

8.11   Conclusion

Despite recent work in techniques for choosing relevant variants and the success that has resulted, there are still numerous ways that machine learning may enhance its research of this crucial issue. The introduction of increasingly complex datasets is urgent. So far, hardly any of the domains investigated have had more than 40 characteristics. The research of cloud categorization by researchers, which utilized 204 characteristics, is an example, although other studies have used only fewer variants. Furthermore, the findings of Langley and Sage's closest neighbor technique indicate that several of the UCI sets of data include few, if any, useless characteristics. In retrospect, this appears reasonable in diagnostic domains, where specialists are more likely to inquire about important characteristics while ignoring others. However, we think that many real-world environments lack this characteristic. Thus if test our variant selection theories properly, we must locate datasets with such a significant proportion of irrelevant characteristics. Experiments using fake data are also useful in researching variant selection techniques. Such datasets allow one to change variables of interest, such as the amount of important and unimportant characteristics, while keeping other variables constant. In this manner, the samples complexity of algorithms may be directly measured as a function of these variables, demonstrating their capacity to scale to areas with many non-essential characteristics. We make a distinction, however, between using fake data for systematic trials and relying on individual fabricated datasets (such as the Monks issues), which seem to be much less helpful. More difficult domains, with more characteristics and a larger percentage of unimportant ones, will need more advanced variant selection techniques. Efficiency improvements would boost the number of states investigated, while such constant factor improvements would reduce the number of states studied. Issues created by exponential growth in the number of variant sets cannot be eliminated. However, looking at these issues through the lens of heuristic search offers some potential answers.

References

1. Aristidis Likasa, Nikos Vlassis, Jakob J. Verbeek,"The global k-means clustering algorithm", the journal of the pattern recognition society, Pattern Recognition 36 (2003) 451–461, 2002.
2. Carlos Ordonez, "Clustering Binary Data Streams with K-means", San Diego, CA, USA. Copyright 2003, ACM 1- 58113-763-x, DMKD'03, June 13, 2003.
3. K. Wang et al., "A Trusted Consensus Scheme for Collaborative Learning in the Edge AI Computing Domain," in IEEE Network, vol. 35, no. 1, pp. 204–210, January/February 2021, doi:https://doi.org/10.1109/MNET.011.2000249.
4. Guha, D. Samanta, A. Banerjee and D. Agarwal, "A Deep Learning Model for Information Loss Prevention From Multi-Page Digital Documents," in IEEE Access, vol. 9, pp. 80451–80465, 2021, doi:https://doi.org/10.1109/ACCESS.2021.3084841.
5. Rohan Kumar, Rajat Kumar, Pinki Kumar, Vishal Kumar, Sanjay Chakraborty, Prediction of Protein-Protein interaction as Carcinogenic using Deep Learning Techniques, 2nd International Conference on Intelligent Computing, Information and Control Systems (ICICCS), Springer, pp. 461–475, 2021.
6. Guha, A., Samanta, D. Hybrid Approach to Document Anomaly Detection: An Application to Facilitate RPA in Title Insurance. Int. J. Autom. Comput. 18, 55–72 (2021). doi:https://doi.org/10.1007/s11633-020-1247-y
7. Lopamudra Dey, Sanjay Chakraborty, Anirban Mukhopadhyay. Machine Learning Techniques for Sequence-based Prediction of Viral-Host Interactions between SARS-CoV-2 and Human Proteins. Biomedical Journal, Elsevier, 2020.
8. Khamparia, A, Singh, PK, Rani, P, Samanta, D, Khanna, A, Bhushan, B. An internet of health things-driven deep learning framework for detection and classification of skin cancer using transfer learning. Trans Emerging Tel Tech. 2020;e3963. doi:https://doi.org/10.1002/ett.3963
9. Jiawei Han and Micheline Kamber, Data Mining concepts and techniques, Morgan Kaufmann (publisher) from chapter-7 'cluster analysis', ISBN:978-1-55860-901-3, 2006.
10. Dunham, M.H., Data Mining: Introductory And Advanced Topics, New Jersey: Prentice Hall, ISBN-13: 9780130888921. 2003.
11. H. Witten, Data mining: practical machine learning tools and techniques with Java implementations San-Francisco, California: Morgan Kaufmann, ISBN: 978-0-12-374856-0 2000.
12. Kantardzic, M. Data Mining: concepts, models, method, and algorithms, New Jersey: IEEE press, ISBN: 978-0-471-22852-3, 2003.
13. Michael K. Ng, Mark Junjie Li, Joshua Zhexue Huang, and Zengyou He, "On the Impact of Dissimilarity Measure in k-Modes Clustering Algorithm", IEEE transaction on pattern analysis and machine intelligence, vol. 29, No. 3, March 2007.
14. Nareshkumar Nagwani and Ashok Bhansali, "An Object Oriented Email Clustering Model Using Weighted Similarities between Emails Attributes", International Journal of Research and Reviews in Computer science (IJRRCS), Vol. 1, No. 2, June 2010.
15. Oyelade, O. J, Oladipupo, O. O, Obagbuwa, I. C, "Application of k-means Clustering algorithm for prediction of Students' Academic Performance", (IJCSIS) International Journal of Computer Science and Information security, Vol. 7, No. 1, 2010.
16. S. Jiang, X. Song, "A clustering based method for unsupervised intrusion detections". Pattern Recognition Letters, PP. 802–810, 2006.

17. Steven Young, Itemer Arel, Thomas P. Karnowski, Derek Rose, University of Tennesee, "A Fast and Stable incremental clustering Algorithm", TN 37996, 7th International 2010.
18. Taoying Li and Yan Chen, "Fuzzy K-means Incremental Clustering Based on K-Center and Vector Quantization", Journal of computers, vol. 5, No. 11, November 2010.
19. Tapas Kanungo, David M. Mount, "An Efficient k-Means Clustering Algorithm: Analysis and implementation IEEE transaction vol. 24 No. 7, July 2002.
20. Zuriana Abu Bakar, Mustafa Mat Deris and Arifah Che Alhadi, "Performance analysis of partitional and incremental clustering", SNATI, ISBN-979-756-061—6, 2005.
21. Xiaoke Su, Yang Lan, Renxia Wan, and Yuming, "A Fast Incremental Clustering Algorithm", international Symposium on Information Processing (ISIP'09), Huangshan, P.R. China, August-21-23, pp: 175–178, 2009.
22. Kehar Singh, Dimple Malik and Naveen Sharma, "Evolving limitations in K-means algorithm in data Mining and their removal", IJCEM International Journal of Computational Engineering & Management, Vol. 12, April 2011.
23. Anil Kumar Tiwari, Lokesh Kumar Sharma, G. Rama Krishna, "Entropy Weighting Genetic k-Means Algorithm for Subspace Clustering", International Journal of Computer Applications (0975– 8887), Volume 7– No. 7, October 2010.
24. K. Mumtaz, Dr. K. Duraiswamy, "An Analysis on Density Based Clustering of Multi Dimensional Spatial Data", Indian Journal of Computer Science and Engineering, Vol. 1 No 1, pp-8–12, ISSN: 0976-5166.
25. A.M. Sowjanya, M. Shashi, "Cluster Feature-Based Incremental Clustering Approach (CFICA) For Numerical Data, IJCSNS International Journal of Computer Science and Network Security, VOL. 10 No. 9, September 2010.
26. Martin Ester, Hans-Peter Kriegel, Jorg Sander, Michael Wimmer, Xiaowei Xu, "Incremental clustering for mining in a data ware housing", 24th VLDB Conference New York, USA, 1998.
27. Sauravjyoti Sarmah, Dhruba K. Bhattacharyya, "An Effective Technique for Clustering Incremental Gene Expression data", IJCSI International Journal of Computer Science Issues, Vol. 7, Issue 3, No 3, May 2010.
28. Debashis Das Chakladar and Sanjay Chakraborty, Multi-target way of cursor movement in brain computer interface using unsupervised learning, Biologically Inspired Cognitive Architectures (Cognitive Systems Research), Elsevier, 2018.
29. Althar, R.R., Samanta, D. The realist approach for evaluation of computational intelligence in software engineering. Innovations Syst Softw Eng 17, 17–27 (2021). doi:https://doi.org/10.1007/s11334-020-00383-2.
30. B. Naik, M. S. Obaidat, J. Nayak, D. Pelusi, P. Vijayakumar and S. H. Islam, "Intelligent Secure Ecosystem Based on Metaheuristic and Functional Link Neural Network for Edge of Things," in IEEE Transactions on Industrial Informatics, vol. 16, no. 3, pp. 1947–1956, March 2020, doi:https://doi.org/10.1109/TII.2019.2920831.
31. Debashis Das Chakladar and Sanjay Chakraborty, EEG Based Emotion Classification using Correlation Based Subset Selection, Biologically Inspired Cognitive Architectures (Cognitive Systems Research), Elsevier, 2018.
32. D. Samanta et al., "Cipher Block Chaining Support Vector Machine for Secured Decentralized Cloud Enabled Intelligent IoT Architecture," in IEEE Access, vol. 9, pp. 98013–98025, 2021, doi:https://doi.org/10.1109/ACCESS.2021.3095297.
33. CHEN Ning, CHEN An, ZHOU Long-xiang, "An Incremental Grid Density-Based Clustering Algorithm", Journal of Software, Vol. 13, No. 1, 2002.
34. Bock, Frederic E., et al. "A Review of the Application of Machine Learning and Data Mining Approaches in Continuum Materials Mechanics." Frontiers in Materials. vol. 6, 2019, p. 110. Frontiers, doi:https://doi.org/10.3389/fmats.2019.00110.
35. Amador, Sandra, et al. "Chapter 6 - Data Mining and Machine Learning Techniques for Early Detection in Autism Spectrum Disorder." Neural Engineering Techniques for Autism Spectrum Disorder, edited by Ayman S. El-Baz and Jasjit S. Suri, Academic Press, 2021, pp. 77–125. ScienceDirect, doi:https://doi.org/10.1016/B978-0-12-822822-7.00006-5.

36. Dabhade, Pranav, et al. "Educational Data Mining for Predicting Students' Academic Performance Using Machine Learning Algorithms." Materials Today: Proceedings, June 2021. ScienceDirect, doi:https://doi.org/10.1016/j.matpr.2021.05.646.
37. Dogan, Alican, and Derya Birant. "Machine Learning and Data Mining in Manufacturing." Expert Systems with Applications, vol. 166, Mar. 2021, p. 114060. ScienceDirect, doi:https://doi.org/10.1016/j.eswa.2020.114060.
38. Emami Javanmard, Majid, et al. "Data Mining with 12 Machine Learning Algorithms for Predict Costs and Carbon Dioxide Emission in Integrated Energy-Water Optimization Model in Buildings." Energy Conversion and Management, vol. 238, June 2021, p. 114153. ScienceDirect, doi:https://doi.org/10.1016/j.enconman.2021.114153.
39. Jimenez-Carvelo, Ana M., and Luis Cuadros-Rodríguez. "Data Mining/Machine Learning Methods in Foodomics." Current Opinion in Food Science, vol. 37, Feb. 2021, pp. 76–82. ScienceDirect, doi:https://doi.org/10.1016/j.cofs.2020.09.008.
40. Lord, Dominique, et al. "Chapter 12 - Data Mining and Machine Learning Techniques." Highway Safety Analytics and Modeling, edited by Dominique Lord et al., Elsevier, 2021, pp. 399–428. ScienceDirect, doi:https://doi.org/10.1016/B978-0-12-816818-9.00016-0.
41. Ma, Ying, et al. "Meta-Analysis of Cellular Toxicity for Graphene via Data-Mining the Literature and Machine Learning." Science of The Total Environment, vol. 793, Nov. 2021, p. 148532. ScienceDirect, doi:https://doi.org/10.1016/j.scitotenv.2021.148532.
42. Yang, Xin-She. "Chapter 16 - Data Mining and Deep Learning." Nature-Inspired Optimization Algorithms (Second Edition), edited by Xin-She Yang, Academic Press, 2021, pp. 239–58. ScienceDirect, doi:https://doi.org/10.1016/B978-0-12-821986-7.00023-8.
43. Zhao, Qingkun, et al. "Machine Learning-Assisted Discovery of Strong and Conductive Cu Alloys: Data Mining from Discarded Experiments and Physical Features." Materials & Design, vol. 197, Jan. 2021, p. 109248. ScienceDirect, doi:https://doi.org/10.1016/j.matdes.2020.109248.
44. Zou, Chengxiong, et al. "Integrating Data Mining and Machine Learning to Discover High-Strength Ductile Titanium Alloys." Acta Materialia, vol. 202, Jan. 2021, pp. 211–21. ScienceDirect, doi:https://doi.org/10.1016/j.actamat.2020.10.056.

Index

Printed in the United States
by Baker & Taylor Publisher Services